D0744352

Modern Semiconductor Design Series

Verilog Designer's Library

Bob Zeidman
The Chalkboard Network, Cupertino CA

Prentice Hall PTR
Upper Saddle River, New Jersey 07458
http://www.phptr.com

ISBN 0-13-081154-8

Library of Congress Cataloging-in-Publication Data

Zeidman, Bob.
    Verilog designer's library / Bob Zeidman.
        p. cm.
    ISBN 0-13-081154-8
        1. Verilog (Computer hardware description language) 2. Electronic
    digital computers—Computer-aided design.   I. Title
    TK7885.7.Z45   1999
      621.39'2—dc21

99-31214
CIP

Acquisitions editor: *Bernard M. Goodwin*
Editorial supervision/page composition: *Scott Disanno*
Cover design director: *Jerry Votta*
Cover design: *Scott Shearer*
Manufacturing manager: *Alan Fischer*
Marketing Manager: *Lisa Konzelmann*
Project coordinator: *Anne Trowbridge*

© 1999 by Prentice Hall PTR
Prentice-Hall, Inc.
Upper Saddle River, New Jersey 07458

Prentice Hall books are widely used by corporations and government agencies for training, marketing, and resale. The publisher offers discounts on this book when ordered in bulk quantities. For more information, contact Corporate Sales Department, Phone: 800-382-3419; FAX: 201-236-7141; E-mail: corpsales@prenhall.com or write Coporate Sales Department, Prentice Hall PTR, One Lake Street, Upper Saddle River, NJ 07458.

All rights reserved. No part of this book may be reproduced, in any form or by any means, without permission in writing from the publisher.

Printed in the United States of America
10 9 8 7 6 5 4
Fourth Printing
ISBN 0-13-081154-8

Prentice-Hall International (UK) Limited, *London*
Prentice-Hall of Australia Pty. Limited, *Sydney*
Prentice-Hall Canada Inc., *Toronto*
Prentice-Hall Hispanoamericana, S.A., *Mexico*
Prentice-Hall of India Private Limited, *New Delhi*
Prentice-Hall of Japan, Inc., *Tokyo*
Prentice-Hall of Asia (*Singapore*) Pte. Ltd.
Editora Prentice-Hall do Brasil, Ltda., *Rio de Janeiro*

In memory of my dad, Morris Zeidman, whose influence on me was immeasurable, and whose confidence in me and encouragement for anything and everything I attempted is still my greatest motivation.

Contents

Foreword

Over the last decade, digital system design has changed enormously. Whereas in the 1980's a designer would draw logic diagrams consisting of simple gates and flip-flops, today's digital designer tends to specify his design at a higher level of abstraction, mainly at the register-transfer level. Traditional tools such as schematic capture software and gate-level simulators have been replaced by hardware description languages driving behavioral simulators and logic synthesis tools. This evolution has increased the productivity of hardware designers by at least an order of magnitude.

In today's design methodology, designers often perform top-down system designs implementing a system as a collection of interconnected functional blocks. These functional blocks can be registers, adders, multipliers, etc. This book deals with the design of these functional blocks using a hardware description language (Verilog HDL) and logic synthesis tools. Most of the functional blocks used by a typical design engineer are described in this book in a generic fashion. In each chapter, one or more examples of designs of a class of functional block are elaborated. By studying the Verilog HDL description of one of these examples, a novice designer can quickly learn about the details of a particular functional block. It should be easy to understand how a functional block works from reading the Verilog HDL description. When some details are not clear, a simulation of the block will quickly enlighten the designer. A more experienced designer should be able to start out with the Verilog HDL description of a functional block, and then to modify it such that it will become a variant of the example better suited to the designer's need. In real design practice, most engineers modify existing designs to speed up digital design considerably.

The examples provided in this book are an excellent starter set of functional descriptions for the basic building blocks needed by most designers. Therefore, this book should appeal both to the novice digital system designers as well as to more advanced and experienced designers looking for examples on how to construct classes of blocks that they are not familiar with. It provides all designers with numerous real-life design details, which should prove invaluable.

Bill vanCleemput
President
Delos Research Group
Sunnyvale, CA

Preface

Hardware Description Languages (HDLs) are fast becoming the design method of choice for electrical engineers. Their ability to model and simulate all levels of design, from abstract algorithms and behavioral functions to register transfer level (RTL) and gate level descriptions, make them extremely powerful tools. Synthesis software allows engineers to take these very high-level descriptions of chips and systems and automatically convert them to real netlists for manufacturing, at least in theory. As chip complexity increases, and gate counts commonly reach 100,000 and above, HDLs become the only practical design method. Even FPGA densities have increased to the point where HDLs are the most efficient design entry method. The benefits of HDLs are even trickling into the areas of PCB design, where it is useful to have one set of tools for simulating integrated circuits and PCBs and the systems into which they are incorporated. The value of using HDLs to model a system on a behavioral level also cannot be ignored as system architects use them to determine and eliminate bottlenecks and improve overall performance of a wide variety of systems.

WHAT IS THIS BOOK ABOUT?

Of the HDLs available, Verilog is one of the most popular. Many designs have been created in Verilog and a large number of Verilog simulators, compilers, synthesizers, and other tools are available from numerous vendors. Its powerful features have led to many applications in all areas of chip design.

This book provides a library of general purpose routines that simplify the task of Verilog programming and enhance existing designs. I have taken input from other designer engineers to make sure that this library covers many of the common functions that a hardware designer is likely to need. Beginning Verilog designers can use these routines as tutorials in order to learn the language or to increase their understanding of it. Experienced Verilog designers can use these routines as a reference and a starting point for real world designs. Rather than redevelop code for common functions, you can simply cut and paste these routines and modify them for your own particular needs. Each routine includes a brief but complete description plus fully documented Verilog code for Behavioral and Register Transfer Level (RTL) implementations. In addition, the Verilog simulation code that was used to verify each hardware module is also included. This code is also available on the enclosed diskette. Feel free to include the Verilog code, royalty-free, in your own designs.

HOW IS THIS BOOK ORGANIZED?

The routines are organized according to functionality. Each chapter addresses a common type of function such as state machines, memory models, or data flow. Each section of a chapter gives an example of code to implement that particular function. Also, successive sections, in general, have increasingly more complex examples. Each function is described using a behavioral model followed by an RTL model. Because behavioral models do not include low level implementation details, they simulate very fast and can be used for quickly evaluating a proposed architecture for a chip or a system. The behavioral models are also useful for creating a simulation environment for your design. The inputs to a chip can be stimulated using behavioral models that might represent something simple, like DRAMs connected to a microprocessor, or something complex like workstations connected to a network. The RTL code, on the other hand, is needed to create real hardware. It is written with synthesis in mind. Despite the sophistication of many synthesis tools, these programs need to make decisions about the gate level implementation based on the RTL code. For this reason, the RTL descriptions must be written in such a way so that there is no ambiguity with respect to what the designer has in mind. Also, the Verilog simulation code is given that is used to test the functionality of each module. This is important because good simulation code will determine whether the hardware will work correctly.

The organization of the book has another advantage. If you are a novice Verilog designer, you can start by studying the simple examples in the beginning and work your way up to the complex examples toward the end. This will give you a very comprehensive understanding of Verilog. If you are an experienced Verilog designer, you can simply jump right to the section that most closely matches your particular design needs. Take that function, play with it, and modify it to suit your design. This will save a significant amount of time by eliminating the need to write the code from scratch.

WHO IS THIS BOOK FOR?

This book is for Verilog users at any level. It assumes a basic familiarity with Verilog structure and syntax. It does not assume any programming background. The book is particularly well suited to hardware designers learning Verilog without having written programs previously, as well as those who have used languages such as BASIC, C, FORTRAN, or Pascal. The book will also appeal to experienced Verilog designers who can skip to the sections that fit their own needs.

This book is valuable to hardware designers, systems analysts, students, teachers, trainers, vendors, system integrators, VARs, OEMs, software developers. and consultants. It is an ideal follow-on or sourcebook for those who have just completed an introductory book or course on Verilog programming.

SUPPORT, SOURCE DISKS, AND COMMENTS

I have simulated all of the examples using SILOS III version 99.115 from Simucad Inc. All of the RTL code has been synthesized using FPGA Express version 3.4 from Synopsys, Inc. As with all published programs, there will surely be last minute changes, which appear on the README.TXT file on the accompanying diskette.

The publisher and I welcome your comments regarding the routines in the book. If you find bugs, discover better ways to accomplish tasks, or can suggest other routines that you think should be included, I am eager to hear about them. To receive notification of revisions and upgrades to the Library, please mail the registration form that appears later in this book.

Bob Zeidman
Cupertino, California
Bob@ZeidmanConsulting.com
www. ZeidmanConsulting.com

Acknowledgments

I would like to thank the following people who helped me as I wrote this book.

Lance Leventhal, who came to me with the idea for this book and helped me prepare a professional and marketable proposal.

Bernard Goodwin, my editor at Prentice-Hall, who encouraged me and supported my book proposal from the beginning.

Jay Michlin at Synopsys who worked so diligently to get all of the necessary contracts signed.

Carlo Treves, for his very thorough reading and critique of this book.

Bill vanCleemput, for his advice and input and for writing the Foreword.

Simon Napper at InnoLogic Systems, for testing the code, finding a bug, and giving me suggestions for fixing it.

I'd also like to thank Carrie Zeidman, my loving wife, who supports me in all of my endeavors no matter how risky or bizarre.

CODING TECHNIQUES

This section describes general techniques for improving your coding for faster simulation or better synthesis of hardware. It also describes important issues to be aware of when writing and simulating code.

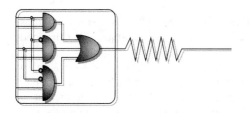

General Coding Techniques

*C*ertain coding practices are simply good practices. Verilog, like other programming languages, offers many features that allow for the creation of very complex structures. However, these features, if used poorly, can create hardware that, while correct, is overly complex or difficult to understand and debug. One important feature of an HDL is the ability to understand the hardware and debug it. In addition, portability is an extremely important consideration. If the code works, but sections cannot be lifted and used for other designs, a critical feature has been lost. In this section, I go over some basic guidelines for producing code that is readable, easy to debug, portable, and modular.

1.1 CODE STRUCTURE

Code should be organized in a way that makes it easy to read and understand, even for newcomers to the design team. In fact, even novice Verilog programmers should be able to pick up the code and understand its function. The structure of the code should be such that the functionality can easily be extended without much rewriting. After all, the main purpose of Verilog, or any HDL, is to allow portability of the code between projects. Writing code that can't be understood defeats this purpose.

I have chosen a template for all of the functions in this book, which is shown below. This template leaves space for each section of Verilog code that might be needed. Even if the code is not needed, the comment is left in. For example, if there are no defines in the code, the // DEFINES comment remains. This is done so that someone examining the code knows exactly where the defines would be. By seeing that the section is blank, the code reviewer knows for certain that there are no defines in this code. He or she does not need to hunt through every line looking for specific keywords.

The header of the template needs to be filled in for each function. The module description comes first. This is followed by the file name, version, date, and author. The code type can be RTL or behavioral. The description should be brief but definitive, so that anyone glancing at the code can know if this is the function that he or she needs. A history of modifications and their dates can be included after the description.

```
/*******************************************************/
// MODULE:        Code template
//
// FILE NAME:     template.v
// VERSION:       1.1
// DATE:          January 1, 2003
// AUTHOR:        Bob Zeidman, Zeidman Consulting
//
// CODE TYPE:     RTL or Behavioral Level
//
// DESCRIPTION:   This template is used to have a coding
// standard for each Verilog function in this book.
//
/*******************************************************/

// DEFINES

// TOP MODULE

// PARAMETERS

// INPUTS

// OUTPUTS

// INOUTS

// SIGNAL DECLARATIONS

// ASSIGN STATEMENTS

// MAIN CODE
```

After the main code, the submodules, tasks, and functions, if any, should be included. They should have their own headers that clearly mark them as a submodule, task, or function. Also, they should have their own comments in a structure similar to the main module, describing the inputs, outputs, etc.

1.2 COMMENTS

The header includes a short description of the function. In addition, the code should include as many comments as possible. I have never seen code with too many comments. I don't think there is such a thing. Of course, the comments should be meaningful. The following comment does nothing to explain the Verilog statement.

```
i = i + 1;        // i is incremented
```

A better comment would be the following.

```
i = i + 1;        // Increment the array index
```

The next example is even better.

```
i = i + 1;        // Since this element of the array did
                  // not meet our criteria, increment the
                  // index to examine the next element
```

1.2.1 Signal Definitions

Each input, output, or inout signal should have a description at the place in the code where it is first defined. An example of this is shown below. Note that groups of signal names and comments read much better if they are lined up exactly at the same place at the page.

```
// INPUTS
input                     reset_n;     // Active low reset
input [`FIFO_WIDTH-1:0]   data_in;     // Data input to FIFO
input                     read_n;      // Read FIFO (active low)
input                     write_n;     // Write FIFO (active low)

// OUTPUTS
output [`FIFO_WIDTH-1:0]  data_out;    // FIFO data output
output                    full;        // FIFO is full
output                    empty;       // FIFO is empty
output                    half;        // FIFO is half full
                                       // or more
```

1.2.2 Code Sections

Each section of code should begin with a general comment describing what it is doing. This includes *always* and *initial* blocks, *if* statements, *while* loops, *case* statements, etc. For each block that includes multiple choices of execution, such as *if* statements, *case* statements, and *fork-join* statements, each alternative should have its own comment. Two examples are shown below.

```verilog
// DESCRIBE THE ALWAYS BLOCK:
// This block contains all devices affected by write input
always @(negedge write_n) begin

    // DESCRIBE THE IF STATEMENT:
    // Check for FIFO overflow
    if (counter >= `FIFO_DEPTH) begin

        // COMMENT FOR THIS BRANCH IS NOT NEEDED
        // BECAUSE THE $DISPLAY STATEMENT IS EXPLANATORY
        $display("\nERROR at time %0t:", $time);
        $display("FIFO Overflow\n");
    end
    else begin

        // COMMENT FOR THIS BRANCH:
        // Store the data
        fifo_mem[wr_pointer] <= #`DEL data_in;
    end
end
```

In the next example, parameters are defined to represent the states of the state machine. In this way, the parameters are used as comments for the states in the *case* statement.

```verilog
// PARAMETERS
parameter[2:0]           // State machine states
    IDLE  = 0,
    WRITE = 3,
    READ1 = 4,
    READ2 = 5;
    .
    .
    .
// Look at the rising edge of clock for state transitions
always @(posedge clock or negedge reset_n) begin
    case (`MEM_STATE)
        IDLE: begin
```

```
            if (wr == 1'b1)
                `MEM_STATE <= #`DEL WRITE;
            else if (rd == 1'b1)
                `MEM_STATE <= #`DEL READ1;
        end
        WRITE: begin
            `MEM_STATE <= #`DEL IDLE;
        end
        READ1: begin
            if (ready == 1'b1)
                `MEM_STATE <= #`DEL READ2;
        end
        READ2: begin
            `MEM_STATE <= #`DEL IDLE;
        end
        default: begin
            `MEM_STATE <= 3'bxxx;
        end
    endcase
end
```

1.2.3 Modules

The end of the each module should have a comment attached that gives the name of the module. This allows the reader to easily match up *end* statements with the corresponding *module* statement. An example is given below. This can also be done for large *begin-end* blocks. The *begin* statement can be labeled and that label can appear as a comment on the *end* statement.

```
module state_machine(
        clock,
        reset_n,
        wr,
        rd,
        ready,
        out_en,
        write_en,
        ack);
    .
    .
    .
endmodule     // state_machine
```

1.3 DO NOT USE *DISABLE* INSTRUCTIONS

The *disable* instruction is a terrible instruction and should not be used. The disable instruction is to Verilog what the *goto* instruction is to the BASIC programming language. In 1968, the famous computer scientist Edsger Dijkstra, wrote a paper entitled "Goto Statement Considered Harmful." At first, the article was met with mixed reactions including some vehement rebuttals. These days, programmers agree that a *goto* instruction in high-level programming languages is not needed and should be avoided at all costs. Why? Let me put this in my own words, because Dijkstra's are very confusing. There are two simple reasons. First, a *goto* statement creates "spaghetti code." That is, code in which modules are intertwined in very non-obvious ways. In spaghetti code, modules cannot be extracted and used in other programs, because other modules may jump to it, or in the worst case, jump into the middle of it. Second, *goto* transfers control in ways that are not obvious. The transfer of control within a section of code can be determined by some instructions that are located in some other module or worse, some other file. This makes understanding the code very difficult. Debugging the code can be a nightmare.

It is for these same exact reasons, that the *disable* statement should not be used. It is true that a *disable* statement can be used to disable the same module that contains the *disable* instruction. This is not as bad as other uses of the instruction. However, other instructions, such as *if-else* and *case* statements, can achieve the same result and are more easily understood when reading code and debugging it. Following are some examples from IEEE Std 1364-1995, "IEEE Standard Hardware Description Language Based on the Verilog® Hardware Description Language," which is the official standard for Verilog. The examples were intended to show how to use the *disable* instruction and the descriptions I use here are taken directly from the standard. I follow each example with my own example showing a better alternative to the *disable* instruction.

1.3.1 Example 1

The following example uses a *disable* instruction in a manner similar to a forward *goto* within a named block. This can also be done within a task.

```
begin: block_name
    .  .  .
    .  .  .
    if (a == 0)
        disable block_name;
    // the rest of the code
    .  .  .
end    // end of named block
// continue with code following named block
.  .  .
```

The next example shows how to accomplish the same thing, end a block, without using a *disable* statement. Again, a task can be terminated in the same way.

```
begin: block_name
    . . .
    . . .
    if (a != 0) begin
        // the rest of the code
        . . .
    end
end    // end of named block
// continue with code following named block
. . .
```

1.3.2 Example 2

The following example shows a *disable* statement used in an equivalent way to the two statements *continue* and *break* in the C programming language. The example illustrates control code that would allow named block to execute until a loop counter reaches n iterations or until the variable a is set to the value of b. The named block block_name contains the code that executes until a == b, at which point the disable block_name; statement terminates execution of that block. The named block loop_name contains the code that executes for each iteration of the *for* loop. Each time the code executes the disable loop_name; statement, the loop_name block terminates and execution passes to the next iteration of the *for* loop. For each iteration of the loop_name block, a set of statements executes if (a != 0). Another set of statements executes if (a != b).

```
begin : block_name
    for (i = 0; i < n; i = i+1) begin : loop_name
        @clk
            if (a == 0)
                disable loop_name;
            // continue if (a != 0)
            . . .
        @clk
            if (a == b)
                disable block_name;
            // continue if (a != b)
            . . .
    end
end
```

The same functionality can be created without the *disable* instruction, using the following code. The (a != b) condition has been put into the *for* loop termination condition. Instead of breaking the loop if (a == 0), we check whether (a != 0) in order to execute the rest of the loop, by putting the code into an *if* statement. In order to execute the next section of code when (a != b) we simply put that code into an *if* statement. Note how much easier this code is to understand.

```
begin
    for (i = 0; (i < n) & (a != b); i = i+1) begin
        @clk
            if (a != 0) begin
                // continue if (a != 0)
                . . .
                @clk
                if (a != b) begin
                    // continue if (a != b)
                    . . .
                end
            end
    end
end
```

1.3.3 Example 3

This example shows the *disable* statement being used to disable concurrently a sequence of timing controls and the task action, when the reset event occurs. The example shows a *fork/join* block within which is a named sequential block (event_expr) and a *disable* statement that waits for the occurrence of the event reset. The sequential block and the wait for reset execute in parallel. The event_expr block waits for one occurrence of event ev1 and three occurrences of event trig. When these four events have happened, plus a delay of d time units, the task action executes. When the event reset occurs, regardless of events within the sequential block, the *fork/join* block terminates–including the task action.

```
fork
    begin : event_expr
        @ev1;
        repeat (3) @trig;
        #d action (areg, breg);
    end
    @reset disable event_expr;
join
```

The following code performs the same function without using the *disable* instruction. In addition, the ac tion task must include code to terminate itself when the reset event occurs. Although this takes slightly more code, it will be obvious by looking at the action task code, that it will terminate when reset occurs.

```
reset_flag = 0;
fork
    begin : event_expr
        @ev1;
        repeat (3) @(trig or posedge reset_flag);
        if (!reset_flag)
            #d action (areg, breg);
    end
    @reset reset_flag = 1;
join
```

1.3.4 Example 4

The last example is a behavioral description of a retriggerable monostable multivibrator (a one-shot). The name event `retrig` restarts the monostable time period. If `retrig` continues to occur within 250 time units, then `q` will remain at `1`.

```
always begin : monostable
    #250 q = 0;
endalways @retrig begin
    disable monostable;
    q = 1;
end
```

Okay, this one really stumped me. I finally found a reference to a solution that is ingenious, because it uses the simulation time as a variable. The code is shown below. The variable `delay_time` contains the value of the most recent time that the retrigger control, `retrig`, has changed. This is done through the first *always* block which stores the time of change of `retrig` in the variable `retrig_time`. The assign statement delays a corresponding change in `delay_time` by 250 time units. The second always block then looks at each change of `delay_time`. If, when `delay_time` changes, `retrig_time` equals `delay_time`, then `delay_time` is exactly 250 time units after the most recent change of `retrig`, and so the one-shot output goes low.

```
module one_shot(
    q,
    retrig);

output      q;
input       retrig;

reg         q;              // one-shot output
wire        retrig;         // control input

reg [63:0]  retrig_time;    // time of the most recent change
                            // of the retrig input
```

```
wire [63:0]   delay_time;        // retrig_time delayed by 250
                                 // time units

assign #250 delay_time = retrig_time;

always @(retrig) begin
    retrig_time = $time;         // save the time of the most
                                 // recent change of retrig
    q = 1;                       // set the output high
end

always @(delay_time) begin
    // if this is true, the last change of retrig
    // was exactly 250 time units ago, so trigger
    // the one shot
    if (retrig_time == delay_time)
        q = 0;
end
endmodule      // one_shot
```

You should note, however, that this is an asynchronous function, which is something thatis frowned upon. It can be very useful for fast behavioral models, but should not be used for RTL models. This technique is used, for example, to design the behavioral model of a Phase Locked Loop (PLL) later on in this book. For an RTL model, both of the above descriptions would result in a poor and unsynthesizable design. The pulse width is dependent on a delay. If it were a synchronous function, the pulse width would depend on a clock period. Then it could easily be designed without using a *disable* instruction. By generating a clock signal, we can use the following code to design a synchronous one-shot.

```
always begin
    #10 clk = ~clk;  // create a periodic clock
end

always @(posedge clk) begin
    if (cnt_reset == 1) begin
        count = 0;
        cnt_reset = 0;
    else if (count < 26)
        count = count + 1;
    else if (count == 25)
        q = 1;
end

always @retrig begin
    cnt_reset = 1;
end
```

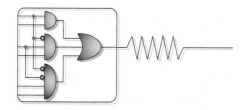

Behavioral Coding Techniques

*I*n general, behavioral models are intended to model the architecture of the device or the algorithm that it realizes, but not the specific implementation. In addition, because behavioral models do not need to be synthesized into hardware, they can use constructs that do not easily represent real hardware. Because these models have fewer constraints, behavioral models should be written to simulate as quickly as possible. The following sections include general structures that work well for behavioral models.

2.1 ELIMINATE PERIODIC INSTRUCTIONS

In order to speed up behavioral model simulations, emphasis should be put on eliminating instructions that are executed periodically. Specifically, any way to reduce the number of instructions that are executed each clock cycle will result in a faster simulation. The most obvious way to do this is by focusing on *always* blocks that are executed on clock edges. By eliminating code within these blocks, the overall simulation time will be reduced. Some examples of this optimization are given in the following sections.

2.1.1 Asynchronous Resets and Presets

For an asynchronous reset or preset, the following code can be implemented. This code forces the output to remain in a particular state whenever the asynchronous input is asserted. Most importantly, the input does not need to be checked during each clock cycle, reducing the number of instructions executed for each clock cycle. This greatly improves the simulation time. Although some synthesis software can recognize this construct as an asynchronous input, and implement it correctly at the gate level, most synthesis software cannot. You should check with your synthesis tool vendor before implementing asynchronous inputs this way in RTL code for synthesis.

```
// Look at the edges of reset
always @(posedge reset or negedge reset) begin
    if (reset)
        assign q = 1'b0;
    else
        deassign q;
end
```

2.1.2 Synchronous Resets, Presets, or Loads

For a synchronous reset, preset, or load, the following code can be implemented. This code forces the output to remain in a particular state whenever the synchronous input is asserted. Again, the input does not need to be checked during each clock cycle, reducing the number of instructions executed for each clock cycle. This greatly improves the simulation time. An *always* block is used to look for changes in the load input. If the load input is asserted, the *always* block is executed, which then looks for the next rising clock edge. If the load input is still asserted at the clock edge, it forces the output to the value of the input. If the load gets deasserted, the *always* block will be executed and the output will immediately be unforced. If, at the next clock edge, the load input has been asserted again, it will once again force the output to the value of the input.

Of course, the assumption here is that our system will be loading data much less often than not, and that the load input will not be changing very often. Also be aware that this code will run slowly if the load input is oscillating due to changes in the combinatorial logic that generates the signal. Even if the signal settles to the correct value before each rising clock, the *always* block will execute whenever the signal changes at least once during a clock cycle. In these cases, this construct should not be used. Instead, the load signal should be checked in the *always* block that depends on the clock signal.

Synthesis software cannot usually recognize this construct. It should be avoided in RTL code. Note the use of a temporary variable, temp, which is simply to store the current input so that the output can be delayed before it is assigned to the value of the input.

```
// Look for the load signal
always @(load) begin
    // if load gets deasserted, unforce the output, but
    // have it keep its current value
    if (~load) begin
        temp = out;
        deassign out;
    end
    // Wait for the rising edge of the clock
    @(posedge clk);

    // If load is asserted at the clock, force the output
    // to the current input value (delayed by `DEL)
    if (load) begin
        temp = in;
        #`DEL assign out = temp;
    end
end
```

2.2 ELIMINATE EVENT ORDER DEPENDENCIES

Verilog was designed to allow the simulation of hardware modules running concurrently. It is important to remember, however, that in reality your Verilog code is being simulated by a sequential processor. This means that events that are supposed to be evaluated simultaneously are actually being evaluated in a particular sequence. In addition, that sequence is unknown to you, and may change for each simulation run. Different simulation software packages can evaluate events in a different order. Even one software package may evaluate simultaneous events in different orders depending on a particular compilation or what the rest of the simulated hardware is doing. In other words, it is impossible to predict which event will be evaluated first.

2.2.1 Use Non-Blocking Assignments

An example of the evaluation problem is shown in the code below.

```
always @(posedge clk) a = b;
always @(posedge clk) b = 0;
```

In one case, a may get set to b, then b is set to 0. This is probably what the designer intended. However, the simulator may reverse the two and set both a and b to 0. To avoid this particular problem, use the non-blocking assignment as shown below, which always will evaluate correctly, regardless of the order of evaluation.

```
always @(posedge clk) a <= b;
always @(posedge clk) b <= 0;
```

2.2.2 Keep Combinatorial Logic Together

Another similar problem is shown in the code below. When b changes, a should immediately change. Will d be set equal to the previous value of a, or the current one?

```
assign a = b & c;
always @(posedge clk) begin
    b = 0;
    d = a;
end
```

Although the designer probably wanted d to take on the new value, it will actually take on the old value because simulators typically evaluate an entire block before reevaluating the assign statements. Where possible, combinatorial logic should be kept in the same block. A better coding of the above logic is shown below.

```
always @(posedge clk) begin
    b = 0;
    d = b & c;
end
```

2.2.3 Use Unit Delays

In order to see more clearly which events depend on which other events, it is often useful to use unit delays for all logic. For example, when we simulate the following code, all of the signals change at the same time and it is difficult to know which ones are dependent on which other ones, without tracing back the code.

```
always @(posedge aclk) begin
    bclk <= 0;
    cclk <= 1;
    dclk <= bclk | cclk;
end
```

However, with the following code, it is easy to see which signals depend on which others because of the unit delay that has been imposed.

```
always @(posedge aclk) begin
    bclk <= #1 0;
    cclk <= #1 1;
    dclk <= #1 bclk | cclk;
end
```

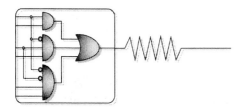

RTL Coding Techniques

*T*his chapter explains general techniques for improved RTL coding. By using these techniques, you will have a much better chance of producing reliable, scalable, portable hardware. First, we must understand exactly what is meant by "reliable, scalable, and portable" with respect to a design, and why these factors are such important considerations.

Reliability seems like it should be an obvious consideration. In my years as a consultant, it has often been the case, unfortunately, that designing for reliability takes a back seat to meeting deadlines or reducing immediate costs. Reliable hardware means that the yield of working hardware will be high. This translates into higher profit margins, which is an obvious good thing. Designing for reliability also will produce a longer mean time between failures (MTBFs) for the hardware, which translates into fewer service calls and lower support costs. But without a good and complete set of rules for reliable design, each and every piece of the design must be examined and analyzed individually for potential problems. Because this is a time consuming process, in companies that do not enforce good design techniques, reliability is seen as a very difficult problem. By following the simple coding techniques described in this chapter, designing for reliability becomes a simple task of following very straightforward guidelines that guarantee reliability.

Scalability is the ability to increase the speed of the process of your design, or decrease the size of the structures of the design, without needing to change the design itself. For an ASIC, this corresponds to speeding up the process by shrinking the structure sizes. By following the guidelines in this chapter, your design will be scalable. This is important for two reasons. First, as technologies speed up, there is competitive pressure to speed up your systems. You don't want your competitor to easily speed up their systems while you are redesigning yours for a higher speed. Second, FPGA vendors are notorious for speeding up their processes without telling their customers. If you look, you will see that FPGA vendors do not specify minimum delays for their processes. When they are able to speed up a semiconductor process and get good yields, they have no reason to continue their old, slow processes. They eventually transfer even their slow devices over to the new process. If your design has marginal timing, you will see problems. This is especially true because not all aspects of the process will speed up proportionally. The switching transistor speed depends on the transistor gate lengths. The trace delay depends on the RC delays of the routing materials used to connect the transistors and the trace geometries. When a process is sped up, the transistors may get faster while the traces delays may remain about the same. Or vice versa. A race condition on your chip, in which one critical signal reaches a device shortly before another critical signal, may now be reversed, completely changing the operation of the circuit as we will see later in this chapter. I have seen this exact situation occur. One company called me in when nearly every controller board in an entire production run of their system failed, even though the board design had not changed for several years. I traced the problem to a race condition in an FPGA. The manufacturer had sped up their FPGA process, causing the chip to fail. This cost the company greatly in terms of lost sales, time tracking the problem, and redesign effort. If you follow the guidelines in this chapter, these types of problems will not happen to you.

Portability is also very important. This is the ability to transfer your design to another vendor and use different software tools to synthesize, simulate, check, and otherwise support the design. By following these guidelines, your design will be portable. This allows you to shop around for the best vendors, even after your design is complete. It also allows you to change manufacturers after you are already in production. This is especially important in today's high tech industry where even today's biggest manufacturers can be out of business tomorrow. And you certainly don't want them to take your design with them to their grave.

3.1 SYNCHRONOUS DESIGN

If there is one commandment for reliable, problem-free design, it is synchronous design. There are essentially two aspects to synchronous design.

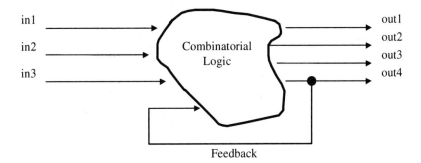

Figure 3-1 Asynchronous logic–combinatorial feedback.

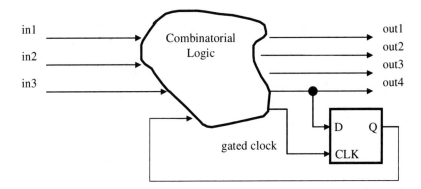

Figure 3-2 Asynchronous logic–gated clock.

No output from combinatorial logic can feed back to affect the same combinatorial logic, without first going through a clocked element.

Clock signals must go directly to the clock inputs of the clocked elements without going through any combinatorial logic.

Figures 3-1 and 3-2 show examples of asynchronous logic. Figure 3-1 has feedback from one block of combinatorial logic that goes directly back to the same block of logic without first passing through a clocked element.

Figure 3-2 shows another example of asynchronous logic. Here, the clock input to the clocked element, a flip-flop, comes from combinatorial logic. This is called a gated clock, and it violates the second rule of synchronous design.

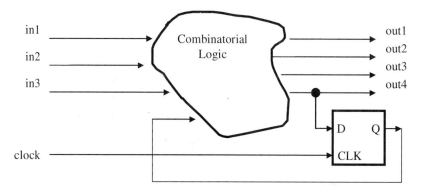

in1

in2

in3

clock

Figure 3-3   Synchronous logic.

Figure 3-3 shows a good example of synchronous logic. The feedback goes through a clocked element and the clock signal goes directly to the clock input of the clocked element, and nowhere else.

The following sections show typical asynchronous design problems and how to avoid them with equivalent synchronous designs.

3.1.1 Race Conditions

Figure 3-4 shows an asynchronous race condition where a clock signal is used to reset a flip-flop. When SIG2 is low, the flip-flop is reset to a low state. On the rising edge of SIG2, the designer wants the output to change to the high state of SIG1. Unfortunately, because we don't know the exact internal timing of the flip-flop or the routing delay of the signal to the clock versus the reset input, we cannot know which signal will arrive first—the clock or the reset. This can be seen in the Verilog code also. It is impossible to predict which `always` block will be executed first. Don't assume that the first one will execute first. The simulation software will do all sorts of optimizations and can easily rearrange the two blocks. If the clock rising edge appears first, the output will remain low. If the reset signal appears first, the output will go high. A slight change in temperature, voltage, or process may cause the race to be won by the wrong signal. A design that always worked correctly will suddenly work incorrectly. A more reliable, synchronous, solution is needed.

The way that I go about finding synchronous solutions is to begin with a state diagram. Figure 3-5 shows the functionality that we need. We start in STATE0, with the output low. When SIG2 is asserted, we check the value of SIG1. If SIG1 is low, we go to STATE1 and keep the output low. If SIG1 is high, we go to STATE2 and set the output high. In both of these states, we wait for SIG2 to be deasserted before returning to STATE0 and setting the output low again. The Verilog code for implementing this state machine is also shown. As with all Verilog state machines, we include a default state to inform the synthesis software that we have considered every possibility.

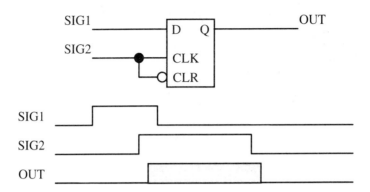

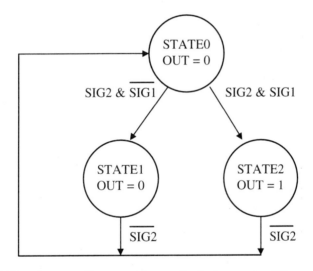

Figure 3-4   Asynchronous: Race condition.

Figure 3-5 Synchronous: State machine to eliminate race condition.

The following code implements the circuit in Figure 3-4.

```
always @(posedge sig2 or negedge sig2) begin
    if (~sig2)
        assign out = 0;
    else
        deassign out;
end

always @(posedge sig2) begin
    out <= sig1;
end
```

The following code implements the circuit in Figure 3-5.

```
always @(posedge clk) begin
    case (state)
        `STATE0: begin
            if (sig2) begin
                if (~sig1)
                    state <= `STATE1;
                else
                    state <= `STATE2;
            end
            else
                state <= `STATE0;
        end
        `STATE1: begin
            if (~sig2)
                state <= `STATE0;
        end
        `STATE2: begin
            if (~sig2)
                state <= `STATE0;
        end
        default: begin
            state <= `STATE0;
        end
    endcase
end
```

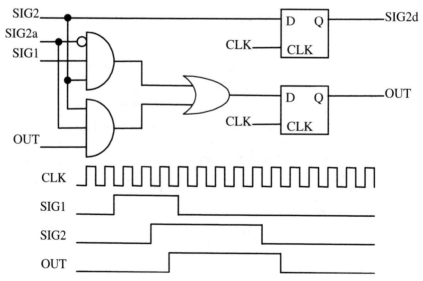

Figure 3-6   Synchronous: No race condition.

The following code implements the circuit in Figure 3-6.

```
always @(posedge clk) begin
    sig2d <= sig;
    out <= (~sig2d & sig1 & sig2) |
        (sig2d & sig2 & out);
end
```

The resulting gate level implementation, after synthesis, will probably look something like Figure 3-6. Here a fast clock is used, and the flip-flop is reset synchronously. This circuit performs the same function as the previous one, but as long as SIG1 and SIG2 are synchronous—they change only after the rising edge of CLK—there is no race condition.

3.1.2 Delay Dependent Logic

Figure 3-7 shows logic used to create a pulse. The pulse width depends very explicitly on the delay of the individual logic gates. If the process should change, making the delay shorter, the pulse width will shorten also, to the point where the logic that it feeds may not recognize it at all. In addition, the synthesis software will most likely optimize the logic so that there is no delay. A synchronous pulse generator is shown in Figure 3-8. This pulse depends only on the clock period. Changes to the process will not cause any significant change in the pulse width.

The following code implements the circuit in Figure 3-7.

```
assign a1 = ~a;
assign a2 = ~a1;
assign a3 = ~a2;
    assign z = a & a3;
```

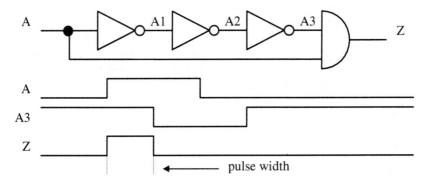

Figure 3-7   Asynchronous: Delay dependent logic.

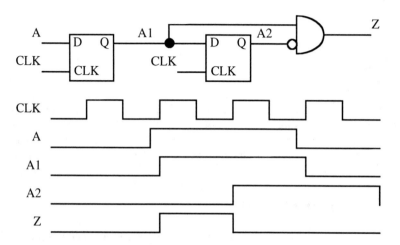

Figure 3-8 Synchronous: Delay independent logic.

The following code implements the circuit in Figure 3-8.

```
always @(posedge clk) begin
    a1 <= a;
    a2 <= a1;
    z <= a1 & ~a2;
end
```

3.1.3 Glitches

A glitch can occur due to small delays in a circuit such as that shown in Figure 3-9. The multiplexer produces a glitch when switching between two signals, both of which are high. Due to the delay of the inverter, the output goes low for a very short time. Synchronizing this output by sending it through a flip-flop as shown in Figure 3-10, ensures that this glitch will not appear on the output and will not affect logic further downstream.

The following code implements the circuit in Figure 3-9.

assign z = sel ? d1 : d0;

The following code implements the circuit in Figure 3-10.

```
always @(posedge clk) begin
    z <= sel ? d1 : d0;
end
```

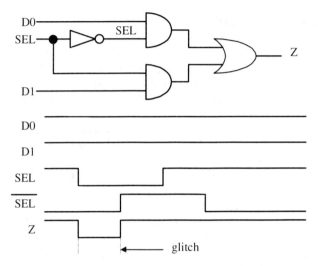

Figure 3-9   Asynchronous: Glitch.

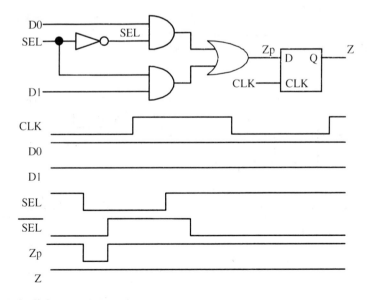

Figure 3-10 Synchronous: No glitch.

3.1.4 Hold Time Violations

Figure 3-11 shows an asynchronous circuit with a hold time violation. Hold time violations occur when data changes around the same time as the clock edge. It is uncertain which value will be registered. Figure 3-12 fixes this problem by putting both flip-flops on the same clock and using a multiplexer to either load new data or keep the previous data.

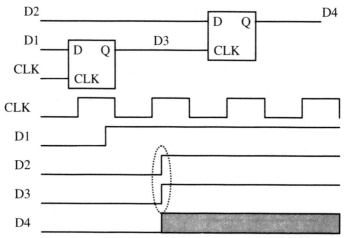

Figure 3-11   Asynchronous: Hold time violation.

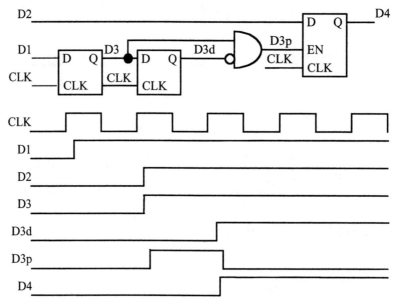

Figure 3-12   Synchronous: No hold time violation.

The following code implements the circuit in Figure 3-11.

```
always @(posedge clk) begin
    d3 <= d1;
end

always @(posedge d3) begin
    d4 <= d2;
end
```

The following code implements the circuit in Figure 3-12.

```
assign d3p = d3 & ~d3d;

always @(posedge clk) begin
    d3 <= d1;
    d3d <= d3;
    d4 <= d3p ? d2 : d4;
end
```

3.1.5 Gated Clocking

Figure 3-13 shows an example of gated clocking. The correct way to enable and disable clocks is not by putting logic on the clock input, but by putting logic on the data input as shown in Figure 3-14. Here a multiplexer is used as a clock enable. When the GATE signal is high, new data is clocked into the flip-flop. When the gate signal is low, the current data is clocked back into the flip-flop.

The following code implements the circuit in Figure 3-13.

```
assign clock1 = gate & clk;

always @(posedge clock1) begin
    out <= data;
end
```

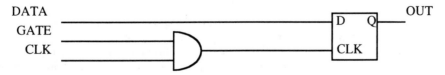

Figure 3-13 Asynchronous: Gated clocking.

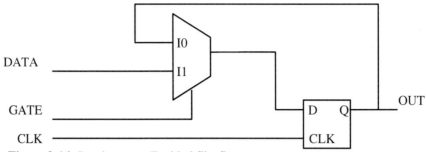

Figure 3-14 Synchronous: Enabled flip-flop.

The following code implements the circuit in Figure 3-14.

```
always @(posedge clk) begin
    out <= gate ? data : out;
end
```

3.1.6 Metastability

One of the great buzzwords, and often misunderstood concepts, of synchronous design is metastability. Metastability refers to a condition that arises when an asynchronous signal is clocked into a synchronous flip-flop. While designers would prefer a completely synchronous world, the unfortunate fact is that signals coming into chip will depend on a user pushing a button or an interrupt from a processor, or can be generated by a clock that is different from the one used by the local logic. In these cases, the asynchronous signal must be synchronized to the local clock so that it can be used. You must be careful to do this in such a way to avoid metastability problems, illustrated in Figure 3-15.

The following code implements the circuit in Figure 3-15.

```
always @(posedge clk) begin
    in <= async_in;
    out1 <= x & in;
    out2 <= x & in;
end
```

If the ASYNC_IN signal goes high around the same time as the clock, we have an unavoidable race condition. The output of the flip-flop can actually go to an undefined voltage level that is somewhere between a logic 0 and logic 1. This is because an internal transistor did not have enough time to fully charge to the correct level. This metalevel may remain until the transistor voltage leaks off or "decays," or until the next clock cycle when a good value is clocked in. During the clock cycle, gates that are connected to output of the flip-flop may interpret this voltage level differently. In the figure, the upper gate sees the level as a logic 1, whereas the lower gate sees it as a logic 0. In normal operation

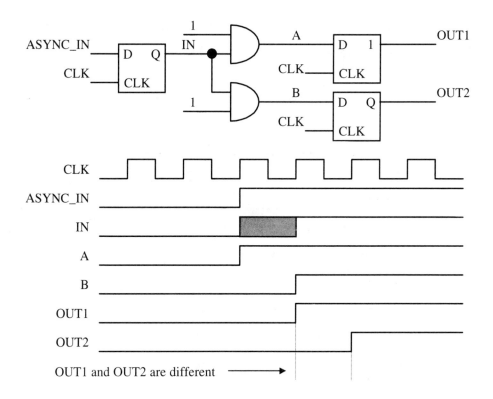

Figure 3-15 Metastability—the problem.

the signals OUT1 and OUT2 should always be the same value. In this case they are not, and this could send the logic into an unexpected state from which it may never return. This metastability condition can permanently lock up your hardware.

The following code implements the circuit in Figure 3-16.

```
always @(posedge clk) begin
    sync_in <= async_in;
    in <= sync_in;
    out1 <= x & in;
    out2 <= x & in;
end
```

The "solution" to this metastability problem is shown in Figure 3-16. By placing a synchronizer flip-flop in front of the logic, the synchronized input will be sampled by only one device, the second flip-flop, and be interpreted only as a logic 0 or 1. The upper and lower gates will both sample the same logic level, and the metastability problem is avoided.

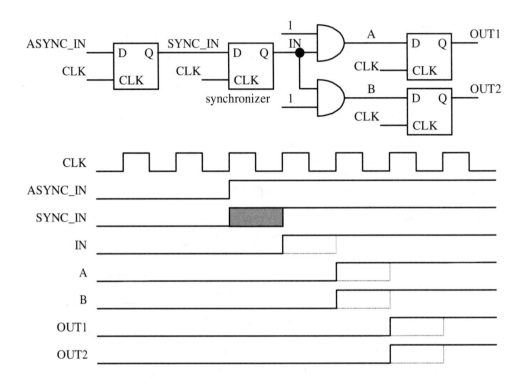

Figure 3-16 Metastability—the "solution."

Or is it? The word "solution" is in quotation marks for a very good reason. There is a very small but non-zero probability that the output of the synchronizer flip-flop will not decay to a valid logic level within one clock period. In this case, the next flip-flop will sample an indeterminate value, and there is again a possibility that the output of that flip-flop will be indeterminate. At higher frequencies, this possibility is greater. Unfortunately, there is no certain solution to this problem. Some vendors provide special synchronizer flip-flops whose output transistors decay very quickly. Also, inserting more synchronizer flip-flops in the chain reduces the probability of metastability, but it will never reduce it to zero. The correct action involves discussing metastability problems with the hardware vendor, and including enough synchronizing flip-flops to reduce the probability so that it is unlikely to occur within the lifetime of the product.

Notice that each synchronizer flip-flop may delay the logic level change on the input by one clock cycle before it is recognized by the internal circuitry. This is shown in Figure 3-16 by the dashed lines on the timing diagram. Given that the external signal is asynchronous, by definition this is not a problem since the exact time that it is asserted will not be deterministic. If this delay is a problem, then most likely the logic that generates this signal, and the circuitry that looks at it, will need to be synchronized to a single clock.

3.2 ALLOWABLE USES OF ASYNCHRONOUS LOGIC

Now that I've gone through a long argument against asynchronous design, I will tell you the few exceptions that I have found to this rule. These exceptions, however, must be designed with extreme caution and only as a last resort when a synchronous solution cannot be found.

3.2.1 Asynchronous Reset

There are times when an asynchronous reset is acceptable, or even preferred. If the vendor's library includes asynchronously resettable flip-flops, the reset input can be tied to a master reset signal in order to reduce the routing congestion and to reduce the logic required for a synchronous reset. FPGA architectures always provide asynchronous resets to their flip-flops. Not using this resource will mean that you will be using up very valuable routing resources for the reset signal. This reset should be used only for resetting the entire hardware and should not occur during normal functioning of the chip. After reset, you must ensure that the design is in a stable state such that no flip-flops will change until an input changes. You must also ensure that the inputs to the hardware are stable and will not change for at least one clock cycle after the reset is removed.

3.2.2 Asynchronous Latches on Inputs

Some buses, such as the VME bus, are designed to be asynchronous. It is usually much more efficient to use asynchronous latches to capture the data initially, and then synchronize the latched data, rather than attempting to synchronize the signals directly. Unless your design uses a clock that has a frequency much higher than that of the bus, attempting to synchronously latch these signals will cause a large amount of overhead and may actually produce timing problems rather than reduce them.

3.2.3 Other Asynchronous Circuits

Occasionally, circuits are needed to operate before a clock has started running or when a clock has stopped running. Circuits that generate system resets, and watchdog circuits are examples of these. Every attempt should be made to design these circuits synchronously or move them off chip and use discrete chips whose worst case and best case timing is very explicitly defined. If this cannot be done, design these circuits with care and realize that changes to the vendor's process may make your chip unusable.

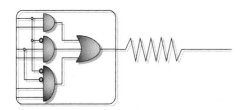

CHAPTER **4**

Synthesis Issues

This chapter gives a description of issues and problems that arise when a synthesis program takes your well-written RTL code and creates gate level code from it. Despite the best efforts of the tool and the programmers that wrote it, the software will need to make certain assumptions that may not match the assumptions that you, the designer, had in mind. Also, due to subtle limitations in the simulation software, the RTL code and the gate level code may not produce identical simulation results. By going over the examples in this chapter, you will be prepared for these types of problems. You will then be able to find and fix the problem quickly, becoming a hero to the design team and a valued employee at your organization.

4.1 CORRELATED UNKNOWN SIGNALS

One problem that has plagued simulation software for a very long time is what is called the problem of correlated unknown signals. To illustrate the problem, let us look at a J-K flip flop shown in Figure 4-1, with its truth table shown in Table 4-1.

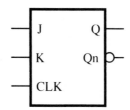

Figure 4-1   A J-K flip flop.

Table 4-1   J-K Flip Flop Truth Table

INPUTS			OUTPUTS	
CLK	**J**	**K**	**Q**	**Qn**
↓	0	0	no change	
↓	1	0	1	0
↓	0	1	0	1
↓	1	1	Toggle	

The RTL code for this flip flop can be found in Chapter 5, with the exception that this flip flop does not have an asynchronous reset. Depending on how the synthesis software decides to implement this function, the synthesized gate-level code could look like the following code.

```
/**********************************************************/
// MODULE:      J-K flip flop
//
// FILE NAME:   jk_gate.v
// VERSION:     1.1
// DATE:        January 1, 2003
// AUTHOR:      Bob Zeidman, Zeidman Consulting
//
// CODE TYPE:   Gate Level
//
// DESCRIPTION: This module defines a J-K flip flop.
//
/**********************************************************/
// DEFINES
`define DEL  1     // Clock-to-output delay. Zero
                   // time delays can be confusing
                   // and sometimes cause problems.
// TOP MODULE
module jk_FlipFlop(
         clk,
         clr_n,
         j,
         k,
         q,
         q_n);
// PARAMETERS
```

```
// INPUTS
input  clk;        // Clock
input  clr_n;      // Asynchronous clear input
input  j;          // J input
input  k;          // K input

// OUTPUTS
output q;          // Output
output q_n;        // Inverted output

// INOUTS

// SIGNAL DECLARATIONS
wire   clk;
wire   clr_n;
wire   j;
wire   k;
reg    q;
wire   q_n;

wire   d1;         // Intermediate gate output
wire   d2;         // Intermediate gate output
wire   d3;         // Intermediate gate output

// ASSIGN STATEMENTS
assign q_n = ~q;
assign d1 = ~k & q;
assign d2 = j & ~q;
assign d3 = d1 | d2;

// MAIN CODE
always @(negedge clk) begin
    q <= #1 d3;
end
endmodule       // jk_FlipFlop
```

The schematic representation of this code is shown in Figure 4-2.

Notice what happens when we try to synchronously preset this circuit after power on. At power on, the outputs Q and Qn are undefined (simulation value X). If we apply a 0 to input K and a 1 to input J, then node D1 goes to X and D2 goes to X, resulting in D3 becoming X. After the clock edge, an undefined value is clocked into the flip flop and the outputs remain undefined. Yet for the RTL description, when simulated, the output Q goes to 1 and Qn goes to 0 as defined in the truth table. It seems that the synthesis software has produced an incorrect gate level implementation.

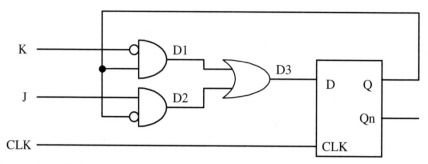

Figure 4-2   Schematic representation of a gate level J-K flip flop.

Actually, the gate level implementation is correct, but the simulator has a limitation. If we look closer at what happens, we see that D1 has the same value as Q while D2 has the same value at Qn. In other words, there is a correlation between the values of these signals, even though they are undefined. Because Q and Qn are inverses of each other, the output of D3 which is the OR of a 1 and a 0 (we just don't know which input is which), must be a 1. Thus, Q will be set to 1 on the next clock edge, as required. Some simulators have the ability to track undefined signals through the circuit as we have just done, but most cannot. Understanding this issue can potentially save many hours debugging post-synthesis simulations. This is another reason why asynchronous resets, if used according to the guidelines given in Chapter 3, can be an advantage to initialize flip-flops in your design.

4.2 STATE MACHINES

When writing Verilog code for state machines, it is important to give the synthesis tool enough information to be able to optimize the state machine properly. In some cases, without the correct information, the synthesizer will produce a bloated state machine. In other cases, it will produce an incorrect one.

Some synthesis software is better at finding and optimizing state machines than others. However, the following rules apply to all tools and will help all of them produce correct, optimal implementations. Note that in the following sections, I talk about synthesis directives that are embedded in the Verilog comments. Each particular program may have slightly different versions of these directives, but they all essentially work the same way. The specific examples that I give are not specific to any one synthesis software vendor, and should (hopefully) work with all vendors. If not, consult the vendor's manual for their specific syntax.

4.2.1 Full Case Specification

When the states of a state machine are specified with a *case* statement, each state must be defined, even if the state cannot be reached or the output is a don't care. Otherwise, the syn-

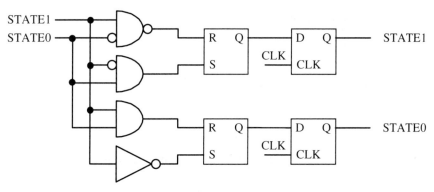

Figure 4-3   Synthesized state machine with latches.

thesis software may assume that the outputs should not change in an undefined state. The way to accomplish this is with a latch, which the software will insert into the resulting RTL code. Latches are not good. At best they result in additional, unnecessary delay. At worst, they create asynchronous timing problems.

In the code below, a state and its outputs are undefined. Figure 4-3 shows the resulting gate level hardware that is produced by the synthesis tool. Note the R-S latches that are put into the design by the synthesis software.

```
// Prepare the next state
always @(state) begin
    case (state)
        2'b00: newstate <= 3'b01;
        2'b01: newstate <= 3'b11;
        2'b11: newstate <= 3'b00;
    endcase
end

// Look at the rising edge of clock for state transitions
always @(posedge clk) begin
    state <= newstate;
end
```

There are two solutions to this problem. The simplest, is to simply put a default in the *case* statement that sets the next state to an undefined value. This tells the synthesis tool that it can optimize all undefined states whichever way is best. This solution is shown below.

```
// Prepare the next state
always @(state) begin
    case (state)
        2'b00: newstate <= 3'b01;
        2'b01: newstate <= 3'b11;
```

```
        2'b11: newstate <= 3'b00;
        default:  newstate <= 3'bx;
    endcase
end

// Look at the rising edge of clock for state transitions
always @(posedge clk) begin
    state <= newstate;
end
```

The other solution is to use the full case directive that is a comment of the form

```
// synthesis full_case
```

on the CASE statement. This is shown in the code below.

```
// Prepare the next state
always @(state) begin
    case (state)  // synthesis full_case
        2'b00: newstate <= 3'b01;
        2'b01: newstate <= 3'b11;
        2'b11: newstate <= 3'b00;
    endcase
end

// Look at the rising edge of clock for state transitions
always @(posedge clk) begin
    state <= newstate;
end
```

Although both method produce correct results, the first method is preferred, because it is standard Verilog and will be recognized by any synthesis tool. Also, it ensures that the simulation will produce the same results for the RTL code as for the synthesized gate level code. The resulting gate level hardware, with no latches, is shown in Figure 4-4. Notice how much simpler this implementation is.

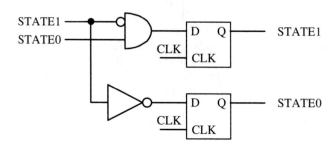

Figure 4-4   Synthesized state machine with no latches.

4.2.2 Parallel Case Specification

Another important way of coding state machines is to make sure that all of the statements in a *case* statement are mutually exclusive. In other words, no two cases can be true at the same time. In Verilog, the *case* statement is defined like a series of *if-else* statements, with priority encoding. This means that the first case is evaluated. If it is true, then no more evaluations take place. If it is false, the next case is evaluated. This continues until a true case is found or all of the cases have been evaluated. It is possible, then, for two cases to be true. Only the first case in the sequence will be evaluated. The problem with this, is that synthesis software has a difficult time optimizing this type of construct in hardware. You specify that the cases are mutually exclusive by putting the comment

```
// synthesis parallel_case
```

on the *case* statement. By specifying parallel case, you are informing the synthesis software that only one case can be true at a time. To see what the resulting hardware is, look at Figures 4-5 and 4-6.

In Figure 4-5, parallel case is not specified, and the synthesis software has included much redundant logic. The synthesis software is unaware that the four select signals are mutually exclusive. Only one will be asserted at any time.

The following code implements the circuit in Figure 4-5.

```
case ({en3,en2,en1})
    3'b??1: out = in1;
    3'b?1?: out = in2;
    3'b1??: out = in3;
endcase
```

The following code implements the circuit in Figure 4-6.

```
case ({en3,en2,en1})// synthesis parallel_case
    3'b??1: out = in1;
    3'b?1?: out = in2;
    3'b1??: out = in3;
endcase
```

In Figure 4-6, parallel case is specified, and the resulting hardware is reduced. It is very important to note that parallel case should only be specified if it is truly the situation that all cases are mutually exclusive. Otherwise, the resulting synthesized logic will be incorrect.

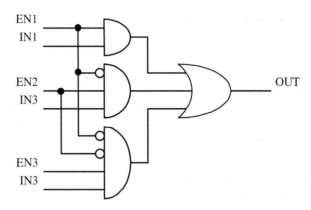

Figure 4-5   CASE statement without parallel case directive.

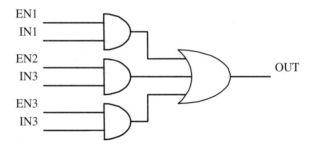

Figure 4-6   CASE statement with parallel case directive.

4.3 OPTIMIZING OUT TERMS

Sometimes it is important to use redundant terms in logic in order to prevent glitches. One example is shown below. One of two signals is used to drive a three-state bus. There are two mutually exclusive selects for the drivers. In order to reduce the time during which both drivers are turned on, the contention time, each driver is driven by ANDing the correct select with the negation of the other select. This turns on one driver only after the other driver has been turned off. The intended logic is shown in Figure 4-7.

Unfortunately, the synthesis tool will see this as redundant logic and eliminate it as shown in Figure 4-8. If this logic needs to be included in the design, the synthesis optimization must be turned off for this block. The method for turning off the optimization is dependent on the particular tool, and you should consult the manual or talk with the vendor.

```
assign sel1 = select ? 1 : 0;
assign sel0 = select ? 0 : 1;
assign en1 = sel1 & ~sel0;
assign en0 = ~sel1 & sel0;

bufif0 buffer0(out, in0, en0);
bufif0 buffer1(out, in1, en1);
```

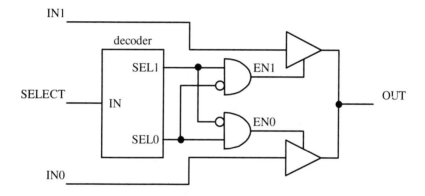

Figure 4-7   Bus select logic before synthesis.

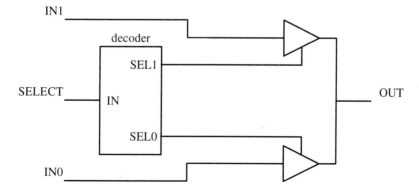

Figure 4-8   Bus select logic after synthesis.

4.4 ALWAYS BLOCKS

An important fact to remember is that synthesis software uses hints from your code to determine ways to optimize your design. One way to help the synthesis software is to combine functions that belong together physically into the same *always* blocks if possible.

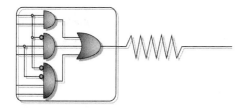

Simulation Issues

*T*his chapter gives a description of issues and problems relating to simulation. Determining that your circuit will work correctly depends on thorough simulation. Simulation must find the majority of functional problems with your circuit before it is committed to hardware. It is extremely important that your simulation code stresses the hardware so that all potential bugs can be determined as early as possible.

5.1 SIMULATE THE CORNER CASES

An important concept in simulation is to simulate the corner cases. This means that the simulation should test conditions that will be difficult to control in a real situation because they occur when the hardware reaches certain extremes. These corner cases include empty memories, full memories, underflowing FIFOs, overflowing FIFOs, asynchronous inputs, etc. The designer should look at the design and decide which combinations of events could cause a failure. Also, it is a good idea to bring in others who are unfamiliar with the design to suggest corner cases for simulation. Many times, the designer has made certain assumptions that outsiders may not make, and that can result in a hardware failure.

5.2 USE CODE COVERAGE TOOLS

Code coverage tools are gaining in popularity, especially for large, complex designs. A code coverage tool will examine your simulations and flag any Verilog statements that have not been executed under all possible conditions by any simulation code. Every statement should, ideally, be executed for each possible condition during some simulation. This gives a certain confidence that there is no unreachable hardware, and that all hardware has been exercised.

5.3 USE THE TRIPLE EQUALS

Probably the most common mistake in writing simulation code has to do with the use of the double equals (==) and the triple equals (===) comparison operator. Because of the use of undefined values and high impedance values, Verilog uses two kinds of tests for equality—logical equality and case equality. Logical equality, the double equals, tests the equality of two well-defined signals. If one signal is not defined as a 1 or 0, then the result of the equality is undefined. Case equality, the triple equals, is needed for simulation and debugging, where you may need to know whether a signal is undefined or high impedance. The logic tables for these operators are shown below.

Table 5-1   Logical Equality

==	0	1	x	z
0	1	0	x	x
1	0	1	x	x
x	x	x	x	x
z	x	x	x	x

Table 5-2   Case Equality

===	0	1	x	z
0	1	0	0	0
1	0	1	0	0
x	0	0	1	0
z	0	0	0	1

In the following code, if `var` is undefined, the *if* statement will not be executed since the condition does not evaluate as true (1).

```
// This will not work
if (var == 1'bx) begin
    // Undefined value found - stop the simulation
    $display("var is undefined!");
    $stop;
end
```

The following code corrects this problem by using the triple equals.

```
// This is correct
if (var === 1'bx) begin
    // Undefined value found - stop the simulation
    $display("var is undefined!");
    $stop;
end
```

For simulation, always use the case equality comparison operator. This ensures that undefined values will not go unnoticed, as they will with a logical equality comparison operator.

5.4 USE THE *$DISPLAY* AND *$STOP* STATEMENTS

The *$display* statement should be used extensively throughout your Behavioral Level and RTL code to give information about the status of the device during simulation. Specifically, error conditions should be displayed to the user and the *$stop* statement should be used when significant functional errors are encountered. These statements allow the hardware code to perform self-checking. These statements are ignored by synthesis software and so they do not change the actual gate level implementation of the hardware.

BASIC BUILDING BLOCKS

This section describes some very basic functions that can be used in almost any design. These functions are also useful as tutorials for basic concepts, syntax, and structures in Verilog.

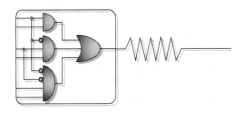

The J-K Flip-Flop

*I*n the days of TTL logic, when a 4-bit microprocessor was considered state of the art, the J-K flip-flop was a very popular device. In particular, the 7473 TTL IC was particularly popular. The reason for this is that it was extremely flexible, allowing one device to act as several different types of flip-flops, depending on how it was wired. This cut down on the number of devices that needed to be purchased and stocked. Also, it meant fewer glue logic devices on a board. In those days, engineers liked to play all sorts of tricks in order to save packages, and the J-K flip-flop lends itself to these tricks.

In fact, when I was younger and not long out of school, I interviewed at a large CAD company for a position as a design engineer. I interviewed with a white-haired, very intense, very intimidating gentleman who gave me a series of very tough technical questions. At one point, he asked me to design a state machine, and I began by drawing some D flip-flops. He immediately began laughing loudly and went into a long lecture about the merits of J-K flip-flops and why no experienced engineer would use anything else. I left demoralized and, needless to say, without a job. I had never used a J-K flip-flop and I assumed that my experience had been lacking. It only occurred to me later that an ASIC designer such as myself had no need for J-K flip-flops. The original advantage of a J-K flip-flop is its high functionality combined with small package and low pin count. But I could throw down any logic I wanted into a chip. D flip-flops have a very simple data path, and are very easy to understand and document, making them the device of choice of ASIC designers. In fact, J-K flip-flops are no longer used at all. However, they do make for a good demonstration of some very basic but important features of Verilog—that is why I include it here.

The J-K flip-flop has two control inputs labeled J and K (hopefully this is not surprising to you). The popular 7473 is clocked on the falling edge of the clock, for some reason unknown to me, but we will design our device that way also. Depending on the J and K inputs, the flip-flop will toggle, set, reset, or clock in new data. The device is shown in Figure 6-1 and the truth table is in Table 6-1. Note that there is an asynchronous clear signal that resets the flip-flop immediately, regardless of the state of the clock.

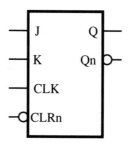

Figure 6-1   A J-K flip-flop.

Table 6-1   J-K Flip-flop Truth Table

INPUTS				OUTPUTS	
CLRn	**CLK**	**J**	**K**	**Q**	**Qn**
0	x	x	x	0	1
1	↓	0	0	no change	
1	↓	1	0	1	0
1	↓	0	1	0	1
1	↓	1	1	toggle	

6.1 BEHAVIORAL CODE

The behavioral code for the implementation of a J-K flip-flop is shown below. Note that the asynchronous clear input is implemented such that once the clear is asserted, the output is forced low until the clear is deasserted. This eliminates the need to test the clear input on each clock cycle during the next always block, reducing the number of instructions that need to be executed each clock cycle. Anything that reduces instructions per clock cycle will speed up the simulation significantly. This works fine for behavioral code, and some synthesis tools can recognize this construct, but most don't. Therefore, in the synthesizable RTL code in the following section, the clear input must be tested during each clock cycle.

```verilog
/***********************************************************/
// MODULE:        j-k flip-flop
//
// FILE NAME:     jk_beh.v
// VERSION:       1.1
// DATE:          January 1, 2003
// AUTHOR:        Bob Zeidman, Zeidman Consulting
//
// CODE TYPE:     Behavioral Level
//
// DESCRIPTION:  This module defines a J-K flip-flop with
// an asynchronous, active low reset.
//
/***********************************************************/

// DEFINES
`define DEL  1      // Clock-to-output delay. Zero
                    // time delays can be confusing
                    // and sometimes cause problems.

// TOP MODULE
module jk_FlipFlop(
          clk,
          clr_n,
          j,
          k,
          q,
          q_n);

// PARAMETERS
// Define the J-K input combinations as parameters
parameter[1:0]   HOLD    = 0,
                 RESET   = 1,
                 SET     = 2,
                 TOGGLE  = 3;
```

```verilog
// INPUTS
input  clk;        // Clock
input  clr_n;      // Active low, asynchronous reset
input  j;          // J input
input  k;          // K input


// OUTPUTS
output q;          // Output
output q_n;        // Inverted output

// INOUTS
// SIGNAL DECLARATIONS
wire   clk;
wire   clr_n;
wire   j;
wire   k;
reg    q;
wire   q_n;


// ASSIGN STATEMENTS
assign #`DEL q_n = ~q;

// MAIN CODE

// Look at the edges of clr_n
always @(posedge clr_n or negedge clr_n) begin
    if (~clr_n)
        #`DEL assign q = 1'b0;
    else
        #`DEL deassign q;
end

// Look at the falling edge of clock for state transitions
always @(negedge clk) begin
    case ({j,k})
        RESET:    q <= #`DEL 1'b0;

        SET:      q <= #`DEL 1'b1;

        TOGGLE:   q <= #`DEL ~q;
    endcase
end
endmodule      // jk_FlipFlop
```

6.2 RTL CODE

The RTL code for the implementation of a J-K flip-flop is shown below. Note that the asynchronous clear signal is implemented such that even after the clear is asserted, the always block is evaluated on each falling clock edge. This increases the simulation time, but allows the synthesis tool to see that both falling edges of the clock and the clear cause immediate changes to the state of the hardware.

There is a potential problem with simulating a synthesized, gate-level J-K flip-flop and many devices like it. The problem can cause undefined signals to propagate throughout the design. For an explanation of the problem, see Chapter 3 regarding correlated unknown signals.

```
/************************************************************/
// MODULE:        j-k flip-flop
//
// FILE NAME:     jk_rtl.v
// VERSION:       1.1
// DATE:          January 1, 2003
// AUTHOR:        Bob Zeidman, Zeidman Consulting
//
// CODE TYPE:     Register Transfer Level
//
// DESCRIPTION:  This module defines a J-K flip-flop with
// an asynchronous, active low reset.
//
/************************************************************/

// DEFINES
`define DEL  1      // Clock-to-output delay. Zero
                    // time delays can be confusing
                    // and sometimes cause problems.

// TOP MODULE
module jk_FlipFlop(
        clk,
        clr_n,
        j,
        k,
        q,
        q_n);

// PARAMETERS
// Define the J-K input combinations as parameters
parameter[1:0]  HOLD      = 0,
                RESET     = 1,
                SET       = 2,
                TOGGLE    = 3;
```

```verilog
// INPUTS
input  clk;       // Clock
input  clr_n;     // Active low, asynchronous reset
input  j;         // J input
input  k;         // K input

// OUTPUTS
output q;         // Output
output q_n;       // Inverted output

// INOUTS

// SIGNAL DECLARATIONS
wire   clk;
wire   clr_n;
wire   j;
wire   k;
reg    q;
wire   q_n;

// ASSIGN STATEMENTS
assign #`DEL q_n = ~q;

// MAIN CODE

// Look at the falling edge of clock for state transitions
always @(negedge clk or negedge clr_n) begin
    if (~clr_n) begin
        // This is the reset condition. Most synthesis tools
        // require an asynchronous reset to be defined this
        // way.
        q <= #`DEL 1'b0;
    end
    else begin
        case ({j,k})
            RESET:    q <= #`DEL 1'b0;

            SET:      q <= #`DEL 1'b1;

            TOGGLE:   q <= #`DEL ~q;
        endcase
    end
end
endmodule    // jk_FlipFlop
```

6.3 SIMULATION CODE

The code used to simulate the J-K flip-flop is shown below. It simply tests each combination of J and K inputs, and the asynchronous clear input. Note that the hold function is tested twice to ensure that the flip-flop correctly holds both 0 and 1 states of the output. The variable `cycle_count` is used to count clock cycles. The *case* statement is then used to check outputs during each clock cycle, and to set up the inputs for the next clock cycle. Because the J-K flip-flop is triggered on the falling edge of the clock, the simulation looks at rising edges of the clock. This allows one half clock cycle before the falling edge to set up inputs, and one half clock cycle after the falling edge to check outputs.

```verilog
/**********************************************************/
// MODULE:          j-k flip-flop simulation
//
// FILE NAME:       jk_sim.v
// VERSION:         1.1
// DATE:            January 1, 2003
// AUTHOR:          Bob Zeidman, Zeidman Consulting
//
// CODE TYPE:       Simulation
//
// DESCRIPTION:  This module provides stimuli for simulating
// a J-K flip-flop. It tests each combination of J and K
// inputs, and the asynchronous clear input. The hold
// function is tested twice to ensure that the flip-flop
// correctly holds both 0 and 1 states of the output.
//
/**********************************************************/
// DEFINES
`define DEL  1       // Clock-to-output delay. Zero
                     // time delays can be confusing
                     // and sometimes cause problems.

// TOP MODULE
module jk_sim();

// PARAMETERS
// Define the J-K input combinations as parameters
parameter[1:0]  HOLD    = 0,
                RESET   = 1,
                SET     = 2,
                TOGGLE  = 3;


// INPUTS

// OUTPUTS
```

```verilog
// INOUTS

// SIGNAL DECLARATIONS
reg       clock;
reg       clear_n;
reg       j_in;
reg       k_in;
wire      out;
wire      out_n;

integer   cycle_count;  // Clock count variable

// ASSIGN STATEMENTS

// MAIN CODE

// Instantiate the flip-flop
jk_FlipFlop jk(
        .clk(clock),
        .clr_n(clear_n),
        .j(j_in),
        .k(k_in),
        .q(out),
        .q_n(out_n));

// Initialize inputs
initial begin
   clock = 0;
   clear_n = 1;
   cycle_count = 0;
end

// Generate the clock
always #100 clock = ~clock;

// Simulate
always @(posedge clock) begin
   case (cycle_count)
      0: begin
         clear_n = 0;
         // Wait for the outputs to change asynchronously
         #`DEL
         #`DEL
         // Test outputs
         if ((out === 0) && (out_n === 1))
            $display ("Clear is working");
         else begin
            $display("\nERROR at time %0t:", $time);
            $display("Clear is not working");
```

```
            $display("    out = %h, out_n = %h\n", out,
                            out_n);
            // Use $stop for debugging
            $stop;
        end
    end
1:  begin
    // Deassert the clear signal
    clear_n = 1;

    {j_in, k_in} = TOGGLE;
    end
2:  begin
    // Test outputs
    if ((out === 1) && (out_n === 0))
        $display ("Toggle is working");
    else begin
        $display("\nERROR at time %0t:", $time);
        $display("Toggle is not working");
        $display("    out = %h, out_n = %h\n", out,
                        out_n);
        // Use $stop for debugging
        $stop;
    end
    {j_in, k_in} = RESET;
    end
3:  begin
    // Test outputs
    if ((out === 0) && (out_n === 1))
        $display ("Reset is working");
    else begin
        $display("\nERROR at time %0t:", $time);
        $display("Reset is not working");
        $display("    out = %h, out_n = %h\b", out,
                        out_n);
        // Use $stop for debugging
        $stop;
    end
    {j_in, k_in} = HOLD;
    end
4:  begin
    // Test outputs
    if ((out === 0) && (out_n === 1))
        $display ("Hold is working");
    else begin
        $display("\nERROR at time %0t:", $time);
        $display("Hold is not working");
        $display("    out = %h, out_n = %h\n", out,
                        out_n);
```

```
                // Use $stop for debugging
                $stop;
            end
            {j_in, k_in} = SET;
        end
        5: begin
            // Test outputs
            if ((out === 1) && (out_n === 0))
                $display ("Set is working");
            else begin
                $display("\nERROR at time %0t:", $time);
                $display("Set is not working");
                $display("    out = %h, out_n = %h\n", out,
                              out_n);
                // Use $stop for debugging
                $stop;
            end
            {j_in, k_in} = HOLD;
        end
        6: begin
            // Test outputs
            if ((out === 1) && (out_n === 0))
                $display ("Hold is working");
            else begin
                $display("\nERROR at time %0t:", $time);
                $display("Hold is not working");
                $display("    out = %h, out_n = %h\n", out,
                              out_n);
                // Use $stop for debugging
                $stop;
            end
            {j_in, k_in} = SET;
        end
        7: begin
            $display("\nSimulation complete - no errors\n");
            $finish;
        end
    endcase
    cycle_count = cycle_count + 1;
end
endmodule     // jk_sim
```

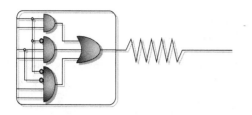

CHAPTER **7**

The Shift Register

A shift register is another common building block for many designs. In this chapter, we look at an 8-bit barrel shifter that can shift any number of bits either left or right. This function can be useful for aligning microprocessor buses of different sizes.

7.1 BEHAVIORAL CODE

The following behavioral code makes use of the fact that the load input takes priority over the other control inputs. An *always* block is used to look for changes in the load input. If the load input is asserted, the *always* block is executed, which then looks for the next rising clock edge. If the load input is still asserted at the clock edge, it forces the output to the value of the input. If the load gets deasserted, the *always* block will be executed and the output will immediately be unforced. If, at the next clock edge, the load input has been asserted again, it will once again force the output to the value of the input. Otherwise, the second *always* block, which is executed on each clock edge, will control the output.

The RTL code in the next section uses a single *always* block for each rising clock edge. The load input is examined along with the other control inputs during this *always* block. The advantage of this behavioral coding is that the *always* block that is executed on each rising clock edge now has fewer instructions. We have sped up execution of this function. Of course, the assumption here is that our system will be loading data into the shift register much less often than it will be idle or shifting data, and that the load input will not be changing very often. If this is not the case, the RTL implementation will execute faster since we use fewer instruction altogether. Also, be aware that this behavioral code will run slower if the load input is oscillating due to changes in the combinatorial logic that generates the signal. Even if the signal settles to the correct value before each rising clock, the first *always* block will execute whenever the signal changes at least once during a clock cycle. Again in this case, the RTL code will run faster and should also be used for the behavioral model.

```
/*************************************************************/
// MODULE:       barrel shifter
//
// FILE NAME:    shift_beh.v
// VERSION:      1.1
// DATE:         January 1, 2003
// AUTHOR:       Bob Zeidman, Zeidman Consulting
//
// CODE TYPE:    Behavioral Level
//
// DESCRIPTION:  This module defines an 8-bit barrel
// shifter.
//
/*************************************************************/

// DEFINES
`define DEL   1      // Clock-to-output delay. Zero
                     // time delays can be confusing
                     // and sometimes cause problems.

// TOP MODULE
module Shifter(
      clk,
      load,
      rshift,
      lshift,
      shiftnum,
      inbit,
      in,
      out);

// PARAMETERS
```

```
// INPUTS
input          clk;          // Clock
input          load;         // Synchronous load
input          rshift;       // Synchronous right shift control
input          lshift;       // Synchronous left shift control
input [2:0]    shiftnum;     // Number of bit places to shift
input          inbit;        // Bit to shift into empty places
input [7:0]    in;           // Input word

// OUTPUTS
output [7:0]   out;          // Output word

// INOUTS

// SIGNAL DECLARATIONS
wire           clk;
wire           load;
wire           rshift;
wire           lshift;
wire [2:0]     shiftnum;
wire           inbit;
wire [7:0]     in;
reg  [7:0]     out;

reg  [7:0]     temp;         // Temporary storage

// ASSIGN STATEMENTS

// MAIN CODE

// Look for the load signal
always @(load) begin
   // if load gets deasserted, unforce the output, but
   // have it keep its current value
   if (~load) begin
      temp = out;
      deassign out;
   end
   // Wait for the rising edge of the clock
   @(posedge clk);
   // If load is asserted at the clock, force the output
   // to the current input value (delayed by `DEL)
   if (load) begin
      temp = in;
      #`DEL assign out = temp;
   end
end
```

```
// Look at the rising edge of the clock
always @(posedge clk) begin
    if (rshift) begin
        // In this implementation, rshift has priority
        // over lshift
        case (shiftnum)
            3'h0: out <= #`DEL out;
            3'h1: out <= #`DEL {inbit,out[7:1]};
            3'h2: out <= #`DEL {inbit,inbit,out[7:2]};
            3'h3: out <= #`DEL {inbit,inbit,inbit,out[7:3]};
            3'h4: out <= #`DEL {inbit,inbit,inbit,inbit,
                    out[7:4]};
            3'h5: out <= #`DEL {inbit,inbit,inbit,inbit,
                    inbit,out[7:5]};
            3'h6: out <= #`DEL {inbit,inbit,inbit,inbit,
                    inbit,inbit,out[7:6]};
            3'h7: out <= #`DEL {inbit,inbit,inbit,inbit,
                    inbit,inbit,inbit,out[7]};
        endcase
    end
    else if (lshift) begin
        // In this implementation, lshift has lowest priority
        case (shiftnum)
            3'h0: out <= #`DEL out;
            3'h1: out <= #`DEL {out[6:0],inbit};
            3'h2: out <= #`DEL {out[5:0],inbit,inbit};
            3'h3: out <= #`DEL {out[4:0],inbit,inbit,
                    inbit};
            3'h4: out <= #`DEL {out[3:0],inbit,inbit,
                    inbit,inbit};
            3'h5: out <= #`DEL {out[2:0],inbit,inbit,
                    inbit,inbit,inbit};
            3'h6: out <= #`DEL {out[1:0],inbit,inbit,
                    inbit,inbit,inbit,inbit};
            3'h7: out <= #`DEL {out[0],inbit,inbit,inbit,
                    inbit,inbit,inbit,inbit};
        endcase
    end
end
endmodule    // Shifter
```

7.2 RTL CODE

```
/**********************************************************/
// MODULE:      barrel shifter
//
// FILE NAME:   shift_rtl.v
// VERSION:     1.1
```

```
// DATE:         January 1, 2003
// AUTHOR:       Bob Zeidman, Zeidman Consulting
//
// CODE TYPE:    Register Transfer Level
//
// DESCRIPTION:  This module defines an 8-bit barrel
// shifter.
//
/***********************************************************/
// DEFINES
`define DEL   1      // Clock-to-output delay. Zero
                     // time delays can be confusing
                     // and sometimes cause problems.
// TOP MODULE
module Shifter(
      clk,
      load,
      rshift,
      lshift,
      shiftnum,
      inbit,
      in,
      out);


// PARAMETERS

// INPUTS
input            clk;        // Clock
input            load;       // Synchronous load
input            rshift;     // Synchronous right shift control
input            lshift;     // Synchronous left shift control
input [2:0]      shiftnum;   // Number of bit places to shift
input            inbit;      // Bit to shift into empty places
input [7:0]      in;         // Input word

// OUTPUTS
output [7:0]  out;           // output word

// INOUTS

// SIGNAL DECLARATIONS
wire         clk;
wire         load;
wire         rshift;
wire         lshift;
wire [2:0]   shiftnum;
wire         inbit;
wire [7:0]   in;
reg  [7:0]   out;
```

```
// ASSIGN STATEMENTS
// MAIN CODE

// Look at the rising edge of the clock
always @(posedge clk) begin
    if (load) begin
        // In this implementation, load has highest priority
        out <= #`DEL in;
    end
    else if (rshift) begin
        // In this implementation, rshift has priority
        // over lshift
        case (shiftnum)
            3'h0: out <= #`DEL out;
            3'h1: out <= #`DEL {inbit,out[7:1]};
            3'h2: out <= #`DEL {inbit,inbit,out[7:2]};
            3'h3: out <= #`DEL {inbit,inbit,inbit,out[7:3]};
            3'h4: out <= #`DEL {inbit,inbit,inbit,inbit,
                          out[7:4]};
            3'h5: out <= #`DEL {inbit,inbit,inbit,inbit,
                          inbit,out[7:5]};
            3'h6: out <= #`DEL {inbit,inbit,inbit,inbit,
                          inbit,inbit,out[7:6]};
            3'h7: out <= #`DEL {inbit,inbit,inbit,inbit,
                          inbit,inbit,inbit,out[7]};
        endcase
    end
    else if (lshift) begin
        // In this implementation, lshift has lowest priority
        case (shiftnum)
            3'h0: out <= #`DEL out;
            3'h1: out <= #`DEL {out[6:0],inbit};
            3'h2: out <= #`DEL {out[5:0],inbit,inbit};
            3'h3: out <= #`DEL {out[4:0],inbit,inbit,
                          inbit};
            3'h4: out <= #`DEL {out[3:0],inbit,inbit,
                          inbit,inbit};
            3'h5: out <= #`DEL {out[2:0],inbit,inbit,
                          inbit,inbit,inbit};
            3'h6: out <= #`DEL {out[1:0],inbit,inbit,
                          inbit,inbit,inbit,inbit};
            3'h7: out <= #`DEL {out[0],inbit,inbit,inbit,
                          inbit,inbit,inbit,inbit};
        endcase
    end
end
endmodule    // Shifter
```

7.3 SIMULATION CODE

The simulation code for the shift register simply performs a number of right shifts and left shifts and compares the actual output to the expected output. It also tests that data can be loaded into the shifter, and that a shift of zero places correctly results in no shift at all.

```
/**********************************************************/
// MODULE:      barrel shifter simulation
//
// FILE NAME:   shift_sim.v
// VERSION:     1.1
// DATE:        January 1, 2003
// AUTHOR:      Bob Zeidman, Zeidman Consulting
//
// CODE TYPE:   Simulation
//
// DESCRIPTION: This module provides stimuli for simulating
// a barrel shifter. It performs a number of right shifts and
// left shifts and compares the actual output to the expected
// output. It also tests that data can be loaded into the
// shifter, and that a shift of zero places correctly results
// in no shift at all.
//
/**********************************************************/

// DEFINES

// TOP MODULE
module shift_sim();

// PARAMETERS

// INPUTS

// OUTPUTS

// INOUTS

// SIGNAL DECLARATIONS
reg           clock;
reg           load;
reg           rshift;
reg           lshift;
reg   [2:0]   numbits;
reg           shift_in;
reg   [7:0]   in_data;
wire  [7:0]   out_data;

reg   [7:0]   expected_data;    // Expected data output
reg   [7:0]   previous_data;    // Previous data output
integer       cycle_count;      // Counter for simulation events
```

```
// ASSIGN STATEMENTS

// MAIN CODE

// Instantiate the parity generator/checker
Shifter shifter(
        .clk(clock),
        .load(load),
        .rshift(rshift),
        .lshift(lshift),
        .shiftnum(numbits),
        .inbit(shift_in),
        .in(in_data),
        .out(out_data));

// Initialize inputs
initial begin
    cycle_count = 0;
    clock = 0;
    load = 0;
    rshift = 0;
    lshift = 0;
end

// Generate the clock
always #100 clock = ~clock;

// Simulate
// Set up the inputs on the falling edge of the clock
always @(negedge clock) begin
    // Check the data output against the expected output
    if (out_data !== expected_data) begin
        $display("\nERROR at time %0t:", $time);
        $display("Output data is incorrect");
        $display("    load input     = %b", load);
        $display("    rshift input    = %b", rshift);
        $display("    lshift input    = %b", lshift);
        $display("    numbits input   = %h", numbits);
        $display("    input bit       = %b", shift_in);
        $display("    input data      = %h", in_data);
        $display("    previous output = %h", previous_data);
        $display("    expected output = %h", expected_data);
        $display("    actual output   = %h\n", out_data);
```

```
    // Use $stop for debugging
    $stop;
end

case (cycle_count)
    1: begin              // Load data
        in_data = 8'b10101010;
        expected_data = 8'b10101010;
        load = 1'b1;
    end
    2: begin              // Shift right
        in_data = 8'bx;
        shift_in = 1'b1;
        numbits = 3'h3;
        expected_data = 8'b11110101;
        load = 1'b0;
        rshift = 1'b1;
    end
    3: begin              // Shift right
        shift_in = 1'b0;
        numbits = 3'h1;
        expected_data = 8'b01111010;
    end
    4: begin              // Shift right
        numbits = 3'h0;
        expected_data = 8'b01111010;
    end
    5: begin              // Shift right
        shift_in = 1'b1;
        numbits = 3'h7;
        expected_data = 8'b11111110;
    end
    6: begin              // Shift left
        shift_in = 1'b0;
        rshift = 1'b0;
        lshift = 1'b1;
        numbits = 3'h2;
        expected_data = 8'b11111000;
    end
    7: begin              // Shift left
        shift_in = 1'b1;
        numbits = 3'h5;
        expected_data = 8'b00011111;
    end
    8: begin              // Load data
        in_data = 8'b11000011;
        expected_data = 8'b11000011;
```

```
            load = 1'b1;
        end
    9: begin
            $display("\nSimulation complete - no errors\n");
            $finish;
        end
    endcase

    // Record the previous data output
    previous_data = out_data;

    // Increment the cycle count
    cycle_count = cycle_count + 1;
end
endmodule          //shift_sim
```

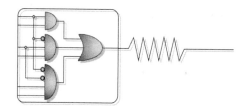

The Counter

*A*nother useful function is the counter. This chapter defines an 8-bit counter with all the bells and whistles. The counter described here has two asynchronous inputs—reset and preset. It has three synchronous inputs—up/down control, count enable, and load enable. There is also a carry output. If there's anything else someone would need in a counter, I can't think of it. The schematic representation of this counter is shown in Figure 8-1.

Note that I have not specified the priority of the different control signals. This is not apparent from the schematic. Here is an area where problems can occur. It is often useful to specify the priorities and test them in the simulation. The asynchronous signals always have top priority over synchronous inputs. If both set and reset are asserted, it is indeterminate which control will prevail, just as will be the case with the real hardware. In this example, I have arbitrarily given the synchronous load input priority over the synchronous count enable. This is evident from the *if-else* statement in the *always* block.

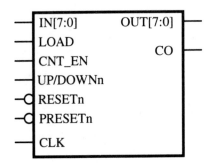

Figure 8-1   An up/down counter.

8.1 BEHAVIORAL CODE

```
/******************************************************/
// MODULE:        up/down counter
//
// FILE NAME:     cnt_beh.v
// VERSION:       1.1
// DATE:          January 1, 2003  03
// AUTHOR:        Bob Zeidman, Zeidman Consulting
//
// CODE TYPE:     Behavioral Level
//
// DESCRIPTION:  This module defines an up/down counter with
// asynchronous set and reset inputs, and synchronous load,
// up/down control, and count enable
//
/******************************************************/

// DEFINES
`define DEL  1        // Clock-to-output delay. Zero
                      // time delays can be confusing
                      // and sometimes cause problems.
`define BITS 8        // Number of bits in counter

// TOP MODULE
module Counter(
      clk,
      in,
      reset_n,
      preset_n,
      load,
```

```
            up_down,
            count_en,
            out,
            carry_out);

// PARAMETERS

// INPUTS
input                   clk;            // Clock
input [`BITS-1:0]       in;             // Input
input                   reset_n;        // Active low,
                                        // asynchronous reset
input                   preset_n;       // Active low,
                                        // asynchronous preset
input                   load;           // Synchronous load input
input                   up_down;        // Synchronous up/down control
input                   count_en;       // Synchronous count
                                        // enable control

// OUTPUTS
output [`BITS-1:0]      out;            // Output
output                  carry_out;         // Carry output

// INOUTS

// SIGNAL DECLARATIONS
wire                    clk;
wire [`BITS-1:0]        in;
wire                    reset_n;
wire                    preset_n;
wire                    load;
wire                    up_down;
wire                    count_en;
reg   [`BITS-1:0]       out;
wire                    carry_out;

// ASSIGN STATEMENTS
// When counting up, the carry output is asserted when all
// outputs are 1. When counting down, it is asserted when all
// outputs are 0.
assign #`DEL carry_out = up_down ? &out : ~(|out);

// MAIN CODE

// Look at the edges of reset
always @(posedge reset_n or negedge reset_n) begin
```

```
    if (~reset_n)
        #`DEL assign out = `BITS'h0;
    else
        #`DEL deassign out;
end

// Look at the edges of preset
always @(posedge preset_n or negedge preset_n) begin
    if (~preset_n)
        #`DEL assign out = ~`BITS'h0;
    else
        #`DEL deassign out;
end

// Look at the rising edge of clock for state transitions
always @(posedge clk) begin
    if (load) begin
        // In this implementation, load has priority over
        // count enable
        out <= #`DEL in;
    end
    else if (count_en) begin
        if (up_down)
            out <= #`DEL out + `BITS'b1;
        else
            out <= #`DEL out - `BITS'b1;
    end
end
endmodule      // Counter
```

8.2 RTL CODE

```
/*************************************************************/
// MODULE:        up/down counter
//
// FILE NAME:     cnt_rtl.v
// VERSION:       1.1
// DATE:          January 1, 2003
// AUTHOR:        Bob Zeidman, Zeidman Consulting
//
// CODE TYPE:     Register Transfer Level
//
```

```
// DESCRIPTION:  This module defines an up/down counter with
// asynchronous set and reset inputs, and synchronous load,
// up/down control, and count enable
//
/*************************************************************/

// DEFINES
`define DEL  1        // Clock-to-output delay. Zero
                      // time delays can be confusing
                      // and sometimes cause problems.
`define BITS 8        // Number of bits in counter

// TOP MODULE
module Counter(
      clk,
      in,
      reset_n,
      preset_n,
      load,
      up_down,
      count_en,
      out,
      carry_out);

// PARAMETERS

// INPUTS
input                 clk;          // Clock
input [`BITS-1:0]     in;           // Input
input                 reset_n;      // Active low,
                                    // asynchronous reset
input                 preset_n;     // Active low,
                                    // asynchronous preset
input                 load;         // Synchronous load input
input                 up_down;      // Synchronous up/down control
input                 count_en;     // Synchronous count
                                    // enable control

// OUTPUTS
output [`BITS-1:0]    out;          // Output
output                carry_out;     // Carry output

// INOUTS

// SIGNAL DECLARATIONS
wire                  clk;
wire [`BITS-1:0]      in;
wire                  reset_n;
```

```
wire                    preset_n;
wire                    load;
wire                    up_down;
wire                    count_en;
reg    [`BITS-1:0]      out;
reg                     carry_up;
reg                     carry_dn;
wire                    carry_out;      // The carry out is generated
                                        // from two registers for
                                        // better clock to output
                                        // timing


// ASSIGN STATEMENTS
assign #`DEL carry_out = up_down ? carry_up : carry_dn;


// MAIN CODE

// Look at the rising edge of clock for state transitions
always @(posedge clk or negedge reset_n or negedge preset_n) begin

    carry_up <= #`DEL 1'b0;
    carry_dn <= #`DEL 1'b0;

    if (~reset_n) begin
        // This is the reset condition.
        out <= #`DEL `BITS'h0;
        carry_dn <= #`DEL 1'b1;
    end
    else if (~preset_n) begin
        // This is the preset condition. Note that in this
        // implementation, the reset has priority over preset
        out <= #`DEL ~`BITS'h0;
        carry_up <= #`DEL 1'b1;
    end
    else if (load) begin
        // In this implementation, load has priority over
        // count enable
        out <= #`DEL in;
        if (in == ~`BITS'h0)
            carry_up <= #`DEL 1'b1;
        else if (in == `BITS'h0)
            carry_dn <= #`DEL 1'b1;
    end
    else if (count_en) begin
        if (up_down) begin
            out <= #`DEL out+1;
            if (out == ~`BITS'h1)
                carry_up <= #`DEL 1'b1;
```

```
        end
        else begin
            out <= #`DEL out-1;
            if (out == `BITS'h1)
                carry_dn <= #`DEL 1'b1;
        end
    end
end
endmodule      // Counter
```

8.3 SIMULATION CODE

The variable `cycle_count` is used to count clock cycles. The *case* statement is then used to check outputs during each clock cycle and to set up the inputs for the next clock cycle. Because the counter is triggered on the rising edge of the clock, the simulation looks at falling edges of the clock. This allows one half clock cycle before the rising edge to set up inputs, and one half clock cycle after the rising edge to check outputs. In cycle 0, we test the asynchronous reset, and in cycle 1, we test the asynchronous preset. In cycle 2, we set up the inputs to load a random pattern into the counter. In cycle 3, we check that the pattern has been loaded. We then deassert the load input and check in cycle 4 that the pattern is still loaded. We then count up until the counter overflows. Then we load another pattern and count down until the counter overflows. The code outside the *case* statement tests that the counter is actually counting correctly. The variables `count_test` and `carry_test` hold expected values for the counter output and the carry out, respectively. They are used to compare to the outputs to test for correctness.

```
/*****************************************************************/
// MODULE:      counter simulation
//
// FILE NAME:   cnt_sim.v
// VERSION:     1.1
// DATE:        January 1, 2003
// AUTHOR:      Bob Zeidman, Zeidman Consulting
//
// CODE TYPE:   Simulation
//
// DESCRIPTION: This module provides stimuli for simulating
// an up/down counter. It tests the asynchronous reset and
// preset controls and the synchronous load control. It loads
// a value and counts up until the counter overflows. It then
// loads a new value and counts down until the counter
// overflows. During each cycle, the output is compared to
// the expected output.
//
/*****************************************************************/
```

```
// DEFINES
`define DEL   1              // Clock-to-output delay. Zero
                            // time delays can be confusing
                            // and sometimes cause problems.
`define BITS 8              // Number of bits in counter

`define PATTERN1 `BITS'h1  // Starting data for counting up
`define PATTERN2 `BITS'h3  // Starting data for counting down

// TOP MODULE
module cnt_sim();

// PARAMETERS

// INPUTS

// OUTPUTS

// INOUTS

// SIGNAL DECLARATIONS
reg                 clock;
reg                 reset_n;
reg                 preset_n;
reg                 load;
reg                 up_down;
reg                 count_en;
reg  [`BITS-1:0]    data_in;
wire [`BITS-1:0]    data_out;
wire                overflow;

reg  [`BITS-1:0]    cycle_count;    // Cycle count variable
integer             test_part;         // Which part of the test
                                    // are we doing?
reg  [`BITS-1:0]    count_test;     // Used to compare against
                                    // the counter output
reg                 carry_test;     // Used to compare against
                                    // the carry output
// ASSIGN STATEMENTS

// MAIN CODE
// Instantiate the counter
Counter counter1(
        .clk(clock),
        .in(data_in),
        .reset_n(reset_n),
```

```
                .preset_n(preset_n),
                .load(load),
                .up_down(up_down),
                .count_en(count_en),
                .out(data_out),
                .carry_out(overflow));

// Initialize inputs
initial begin
    clock = 1;
    reset_n = 1;
    preset_n = 1;
    load = 0;
    count_en = 0;
    cycle_count = `BITS'b0;
    test_part = 0;            // We are doing the first part
                             // of the test
end

// Generate the clock
always #100 clock = ~clock;

// Simulate
always @(negedge clock) begin
    case (test_part)
        0: begin
            case (cycle_count)
                `BITS'h0: begin
                    // Assert the reset signal
                    reset_n = 0;

                    // Wait for the outputs to change
                    // asynchronously
                    #`DEL
                    #`DEL
                    // Test outputs
                    if (data_out === `BITS'h0)
                        $display ("Reset is working");
                    else begin
                        $display("\nERROR at time %0t:", $time);
                        $display("Reset is not working");
                        $display("    data_out = %h\n", data_out);

                        // Use $stop for debugging
                        $stop;
                    end
```

```
      // Deassert the reset signal
      reset_n = 1;

      // Set the expected outputs
      count_test = `BITS'h0;
      carry_test = 1'bx;
   end
`BITS'h1: begin
      // Assert the preset signal
      preset_n = 0;

      // Wait for the outputs to change
      // asynchronously
      #`DEL
      #`DEL
      // Test outputs
      if (data_out === ~`BITS'h0)
         $display ("Preset is working");
      else begin
         $display("\nERROR at time %0t:", $time);
         $display("Preset is not working");
         $display("   data_out = %h\n", data_out);

         // Use $stop for debugging
         $stop;
      end

      // Deassert the preset signal
      preset_n = 1;

      // Set the expected outputs
      count_test = ~`BITS'h0;
   end
`BITS'h2: begin
      // Load data into the counter
      data_in = `PATTERN1;
      load = 1'b1;
   end
`BITS'h3: begin
      // Test outputs
      if (data_out === `PATTERN1)
         $display ("Load is working");
      else begin
         $display("\nERROR at time %0t:", $time);
         $display("Load is not working");
         $display("   expected data_out = %h",
                  `PATTERN1);
```

```verilog
                    $display("     actual    data_out = %h\n",
                              data_out);

                    // Use $stop for debugging
                    $stop;
                end

                // Deassert the load enable signal
                load = 1'b0;

                // Set the expected outputs
                count_test = `PATTERN1;
            end
            `BITS'h4: begin
                // Test outputs to see that data was not lost
                if (data_out === `PATTERN1)
                    $display ("Counter hold is working");
                else begin
                    $display("\nERROR at time %0t:", $time);
                    $display("Counter hold is not working");
                    $display("     expected data_out = %h",
                              `PATTERN1);
                    $display("     actual    data_out = %h\n",
                              data_out);

                    // Use $stop for debugging
                    $stop;
                end

                // Count up
                count_en = 1'b1;
                up_down = 1;

                // Set the expected outputs
                count_test = `PATTERN1;
                carry_test = 1'b0;
            end
            ~`BITS'h0:begin
                // Start the second part of the test
                test_part = 1;
            end
        endcase
    end
    1: begin
        case (cycle_count)
            `BITS'h4: begin
                // Load data into the counter
                data_in = `PATTERN2;
```

```verilog
            load = 1'b1;

            // Set the expected outputs
            count_test = `PATTERN1;
        end
    `BITS'h5: begin
        // Test outputs
        if (data_out === `PATTERN2)
            $display ("Load is working");
        else begin
            $display("\nERROR at time %0t:", $time);
            $display("Load is not working");
            $display("    expected data_out = %h",
                        `PATTERN2);
            $display("    actual   data_out = %h\n",
                        data_out);

            // Use $stop for debugging
            $stop;
        end
        // Count down
        count_en = 1'b1;
        up_down = 1'b0;
        load = 1'b0;

        // Set the expected outputs
        count_test = `PATTERN2;
    end
    ~`BITS'h0:begin
        // Start the third part of the test
        test_part = 2;
    end
    endcase
end
2: begin
    case (cycle_count)
        `BITS'h5:   begin
            $display("\nSimulation complete - no errors\n");
            $finish;
        end
    endcase
end
endcase
```

```
    // Test the counter output
    if (data_out !== count_test) begin
        $display("\nERROR at time %0t:", $time);
        $display("Count is incorrect");
        $display("    expected output = %h", count_test);
        $display("    actual   output = %h\n", data_out);

        // Use $stop for debugging
        $stop;
    end

        // Test the overflow if we are counting
        if ((count_en) && (overflow !== carry_test)) begin
        $display("\nERROR at time %0t:", $time);
        $display("Carry out is incorrect");
        $display("    expected carry = %h", carry_test);
        $display("    actual   carry = %h\n", overflow);
        // Use $stop for debugging
        $stop;
    end

    // Determine the expected outputs for the next cycle
    if (up_down === 1'b1) begin
        if (count_en === 1'b1)
            count_test = count_test + `BITS'h1;
        if (count_test === ~`BITS'h0)
            carry_test = 1'b1;
        else
            carry_test = 1'b0;
    end
    else if (up_down === 1'b0) begin
        if (count_en === 1'b1)
            count_test = count_test - `BITS'h1;
        if (count_test === `BITS'h0)
            carry_test = 1'b1;
        else
            carry_test = 1'b0;
    end

    // Increment the cycle counter
    cycle_count = cycle_count + 1;
end
endmodule     // cnt_sim
```

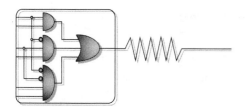

CHAPTER 9

The Adder

*O*ne function that every engineer will need many times in his or her lifetime is an adder. Each particular implementation will be slightly different. In this chapter, I present a 32-bit synchronous adder. Like the other examples in the book, the idea here is to provide a very complex function that can be scaled down and stripped down to fit your particular need. This design can be stripped down for smaller bit widths and less functionality.

The schematic representation of this adder is shown in Figure 9-1. Note that there is an output called valid. This signal is asserted when the output of the adder is valid. This is done because in our behavioral implementation, the output is valid after one clock cycle, whereas the RTL implementation takes eight clock cycles. The reason for this is that the behavioral implementation has no clock cycle timing constraints and can therefore perform the entire calculation in one clock cycle. In the RTL implementation, the clock cycle of the system is an important factor. It is assumed that adding four bits can easily be done within the clock cycle timing requirements of our chip. It may be the case that more

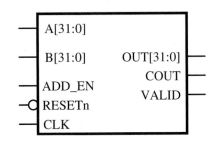

Figure 9-1   An adder.

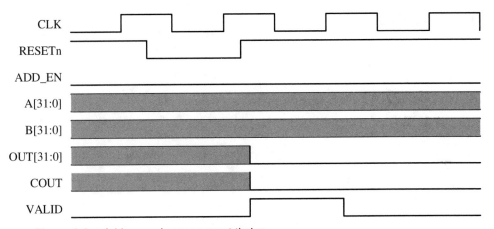

Figure 9-2   Adder synchronous reset timing.

than four bits can be added in one clock cycle. In this case, the RTL code might be modified to use four 8-bit adders, reducing the entire operation to four cycles.

The specific operation of this adder is important, because it has been designed as a pipelined adder. Pipelining allows the additions to occur in stages that can be added or removed, to operate on smaller or larger numbers, without affecting the overall clock cycle time of the system. When the adder is synchronously reset, the outputs go to zero and the valid signal is asserted on the next clock cycle. When the add_en input is asserted and the valid output is asserted during the same clock cycle, the adder begins adding. When the valid output signal is again asserted on a subsequent clock cycle, the new output is correct. Note that the inputs must be held steady from the cycle during which the add enable input is asserted until the cycle during which the valid output signal is asserted. This requirement can be eliminated by latching the most significant bits of the inputs using flip-flops enabled by the add_en input.

The timing for this adder is shown in Figures 9-2 and 9-3. Note that for the behavioral model, because the addition takes place in one clock cycle, the valid signal is always asserted.

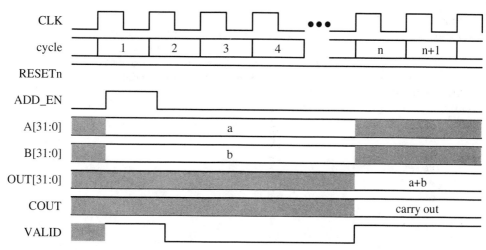

Figure 9-3 Adder timing.

9.1 BEHAVIORAL CODE

```
/************************************************************/
// MODULE:      adder
//
// FILE NAME:   add_beh.v
// VERSION:     1.2
// DATE:        January 1, 2003
// AUTHOR:      Bob Zeidman, Zeidman Consulting
//
// CODE TYPE:   Behavioral Level
//
// DESCRIPTION: This module defines an adder with
// synchronous add enable and reset inputs. When the adder
// is synchronously reset, the outputs go to zero and the
// valid signal is asserted on the next clock cycle. When
// the add enable input is asserted and the valid output is
// asserted during the same clock cycle, the adder begins
// adding. When the valid output signal is again asserted
// on a subsequent clock cycle, the new output is correct.
// Note that the inputs must be held steady from the cycle
// during which the add enable input is asserted until the
// cycle during which the valid output signal is asserted.
//
/************************************************************/
```

```
// DEFINES
`define DEL   1          // Clock-to-output delay. Zero
                         // time delays can be confusing
                         // and sometimes cause problems.
`define BITS 32          // Bit width of the operands
// TOP MODULE
module Adder(
      clk,
      a,
      b,
      reset_n,
      add_en,
      out,
      cout,
      valid);

// PARAMETERS

// INPUTS
input               clk;       // Clock
input [`BITS-1:0]   a;         // Operand A input
input [`BITS-1:0]   b;         // Operand B input
input               reset_n;   // Active low, synchronous reset
input               add_en;    // Synchronous add enable control

// OUTPUTS
output [`BITS-1:0]  out;       // Output
output              cout;      // Carry output
output              valid;     // Is the output valid yet?

// INOUTS

// SIGNAL DECLARATIONS
wire                clk;
wire [`BITS-1:0]    a;
wire [`BITS-1:0]    b;
wire                reset_n;
wire                add_en;
reg  [`BITS-1:0]    out;
reg                 cout;
reg                 valid;

// ASSIGN STATEMENTS

// MAIN CODE
```

```
// Look for the reset_n signal
always @(reset_n) begin
    // if reset gets deasserted, unforce the output
    if (reset_n) begin
        deassign out;
        deassign cout;
        cout = 1'b0;
        out = `BITS'h0;
    end
    // Wait for the rising edge of the clock
    @(posedge clk);
    // If reset is asserted at the clock, force the output
    // to the current input value (delayed by `DEL)
    if (~reset_n) begin
        #`DEL
        assign cout = 1'b0;
        assign out = `BITS'h0;
    end
end

// Look at the rising edge of the clock
always @(posedge clk) begin
    if (add_en)
        {cout, out} <= #`DEL a+b;

    // Output is valid after reset or add operation
    // And remains valid forever
    valid <= #`DEL ~reset_n | add_en | valid;
end
endmodule     // Adder
```

9.2 RTL CODE

```
/*******************************************************/
// MODULE:      adder
//
// FILE NAME:   add_rtl.v
// VERSION:     1.1
// DATE:        January 1, 2003
// AUTHOR:      Bob Zeidman, Zeidman Consulting
//
// CODE TYPE:   Register Transfer Level
//
// DESCRIPTION: This module defines an adder with
// synchronous add enable and reset inputs. When the adder
```

```
// is synchronously reset, the outputs go to zero and the
// valid signal is asserted on the next clock cycle. When
// the add enable input is asserted and the valid output is
// asserted during the same clock cycle, the adder begins
// adding. When the valid output signal is again asserted
// on a subsequent clock cycle, the new output is correct.
// Note that the inputs must be held steady from the cycle
// during which the add enable input is asserted until the
// cycle during which the valid output signal is asserted.
/*************************************************************/

// DEFINES
`define DEL   1        // Clock-to-output delay. Zero
                       // time delays can be confusing
                       // and sometimes cause problems.
// TOP MODULE
module Adder(
      clk,
      a,
      b,
      reset_n,
      add_en,
      out,
      cout,
      valid);

// PARAMETERS

// INPUTS
input            clk;       // Clock
input [31:0]     a;         // 32-bit A input
input [31:0]     b;         // 32-bit B input
input            reset_n;   // Active low, synchronous reset
input            add_en;    // Synchronous add enable control


// OUTPUTS
output [31:0]    out;       // 32-bit output
output           cout;      // Carry output
output           valid;     // Is the output valid yet?


// INOUTS

// SIGNAL DECLARATIONS
wire             clk;
wire [31:0]      a;
wire [31:0]      b;
wire             reset_n;
```

```
wire             add_en;
wire  [31:0]     out;
wire             cout;
wire             valid;
wire  [7:0]      cout4;          // Carry output of 4-bit adder
reg   [2:0]      valid_cnt;         // Counter to determine when the
                            // output is valid


// ASSIGN STATEMENTS
assign #`DEL cout = cout4[7];
assign #`DEL valid = ~|valid_cnt;

// MAIN CODE

// Instantiate eight 4-bit adders
Adder_4bit Add0(
      .clk(clk),
      .a(a[3:0]),
      .b(b[3:0]),
      .cin(1'b0),
      .reset_n(reset_n),
      .add_en(add_en),
      .out(out[3:0]),
      .cout(cout4[0]));

Adder_4bit Add1(
      .clk(clk),
      .a(a[7:4]),
      .b(b[7:4]),
      .cin(cout4[0]),
      .reset_n(reset_n),
      .add_en(add_en),
      .out(out[7:4]),
      .cout(cout4[1]));

Adder_4bit Add2(
      .clk(clk),
      .a(a[11:8]),
      .b(b[11:8]),
      .cin(cout4[1]),
      .reset_n(reset_n),
      .add_en(add_en),
      .out(out[11:8]),
      .cout(cout4[2]));
```

```
Adder_4bit Add3(
     .clk(clk),
     .a(a[15:12]),
     .b(b[15:12]),
     .cin(cout4[2]),
     .reset_n(reset_n),
     .add_en(add_en),
     .out(out[15:12]),
     .cout(cout4[3]));

Adder_4bit Add4(
     .clk(clk),
     .a(a[19:16]),
     .b(b[19:16]),
     .cin(cout4[3]),
     .reset_n(reset_n),
     .add_en(add_en),
     .out(out[19:16]),
     .cout(cout4[4]));

Adder_4bit Add5(
     .clk(clk),
     .a(a[23:20]),
     .b(b[23:20]),
     .cin(cout4[4]),
     .reset_n(reset_n),
     .add_en(add_en),
     .out(out[23:20]),
     .cout(cout4[5]));

Adder_4bit Add6(
     .clk(clk),
     .a(a[27:24]),
     .b(b[27:24]),
     .cin(cout4[5]),
     .reset_n(reset_n),
     .add_en(add_en),
     .out(out[27:24]),
     .cout(cout4[6]));


Adder_4bit Add7(
     .clk(clk),
     .a(a[31:28]),
     .b(b[31:28]),
     .cin(cout4[6]),
     .reset_n(reset_n),
     .add_en(add_en),
```

```
        .out(out[31:28]),
        .cout(cout4[7]));

// Look at the rising edge of the clock
always @(posedge clk) begin
   if (~reset_n) begin
      // Initialize the valid counter
      valid_cnt <= #`DEL 3'h0;
   end
   else if (((valid_cnt == 3'h0) && (add_en == 1'b1)) ||
            (valid_cnt != 3'h0)) begin
      // Increment the valid counter
      // if valid and add_en are asserted
      // or if valid is not asserted
      valid_cnt <= #`DEL valid_cnt + 1;
   end
end
endmodule       // Adder

// SUB MODULE
module Adder_4bit(
     clk,
     a,
     b,
     reset_n,
     add_en,
     cin,
     out,
     cout);

// PARAMETERS

// INPUTS
input           clk;      // Clock
input [3:0]     a;        // 4-bit A input
input [3:0]     b;        // 4-bit B input
input           cin;      // Carry in
input           reset_n;  // Active low, synchronous reset
input           add_en;   // Synchronous add enable control

// OUTPUTS
output [3:0]    out;      // 4-bit output
output          cout;     // Carry output

// INOUTS

// SIGNAL DECLARATIONS
```

```
wire            clk;
wire [3:0]        a;
wire [3:0]        b;
wire            cin;
wire            reset_n;
wire            add_en;
reg  [3:0]        out;
reg             cout;

// ASSIGN STATEMENTS

// MAIN CODE

// Look at the rising edge of the clock
always @(posedge clk) begin
    if (~reset_n) begin
        {cout,out} <= #`DEL 33'h00000000;
    end
    else if (add_en) begin
        {cout,out} <= #`DEL a+b+cin;
    end
end
endmodule        // Adder_4bit
```

9.3 SIMULATION CODE

The code used to simulate the adder is shown below. The variable `cycle_count` is used to count each clock cycles during which `valid` is asserted. In other words, `cycle_count` only counts clock cycles of valid data output. The `val_count` variable counts clock cycles while `valid` is not asserted. Using `val_count`, the simulation can determine if the data output takes too long to become valid, signaling a problem.

The *case* statement is used to check outputs during each valid clock cycle, and to set up the inputs for the next clock cycle. Because the adder is triggered on the rising edge of the clock, the simulation looks at falling edges of the clock. This allows one half clock cycle before the rising edge to set up inputs, and one half clock cycle after the rising edge to check outputs.

In valid cycle 0, we test the synchronous reset input. In valid cycle 1, we deassert the reset input, but we expect the output to remain valid. For the next specified number of valid cycles, we present random data at the inputs to the adder and check that the correct sum and carry out are generated. Finally, we deassert the `add_en` signal and apply new data to the adder inputs. We expect the output to remain unchanged and valid. The *default* condition of the *case* statement simply checks each cycle that is not explicitly defined.

```
/***********************************************************/
// MODULE:          adder simulation
//
// FILE NAME:       add_sim.v
// VERSION:         1.2
// DATE:            January 1, 2003
// AUTHOR:          Bob Zeidman, Zeidman Consulting
//
// CODE TYPE:       Simulation
//
// DESCRIPTION:  This module provides stimuli for simulating
// an adder. The synchronous reset is first checked. Then,
// randomly generated operands are added together and the
// output is compared to the expected sum and carry out.
//
/***********************************************************/

// DEFINES
`define TEST_NUM 255      // Number of addition operations
                          // to test
`define BITS 32           // Bit width of the operands
`define EBITS (`BITS+1)   // Bit width of data plus carry bit

// TOP MODULE
module add_sim();

// PARAMETERS

// INPUTS

// OUTPUTS

// INOUTS

// SIGNAL DECLARATIONS
reg                 clock;
reg                 reset_n;
reg                 add_en;
reg  [`BITS-1:0]    a_in;
reg  [`BITS-1:0]    b_in;
wire [`BITS-1:0]    sum_out;
wire                overflow;
wire                valid;

integer             cycle_count;    // Counts valid clock cycles
integer             val_count;      // Counts cycles between
```

```verilog
                                          // valid data
reg  [`EBITS-1:0]    expect;              // Expected adder output

// ASSIGN STATEMENTS

// MAIN CODE

// Instantiate the adder
Adder adder1(
          .clk(clock),
          .a(a_in),
          .b(b_in),
          .reset_n(reset_n),
          .add_en(add_en),
          .out(sum_out),
          .cout(overflow),
          .valid(valid));

// Initialize inputs
initial begin
    clock = 0;
    reset_n = 0;
    add_en = 1;
    cycle_count = 0;
    val_count = 0;
    a_in = $random(0);      // Initialize random number generator
end

// Generate the clock
always #100 clock = ~clock;

// Simulate
always @(negedge clock) begin
    if (valid === 1'b1) begin
        case (cycle_count)
            0: begin
                // Test outputs
                if ((sum_out === `BITS'h00000000) &&
                    (overflow === 1'b0))
                   $display ("Reset is working");
                else begin
                   $display("\nERROR at time %0t:", $time);
                   $display("Reset is not working");
                   $display("    sum_out = %h", sum_out);
                   $display("    overflow = %b\n", overflow);

                   // Use $stop for debugging
                   $stop;
```

```
      end
      // Test valid output signal
      if (valid === 1'b1)
         $display ("Valid signal is working");
      else begin
         $display("\nERROR at time %0t:", $time);
         $display("Valid signal is not working");
         $display("    valid = %b\n", valid);

         // Use $stop for debugging
         $stop;
      end
      // Deassert the reset signal
      reset_n = 1;
      // Create random inputs
      // between 0 and all 1s
      a_in = {$random} % (`EBITS'h1 << `BITS);
      b_in = {$random} % (`EBITS'h1 << `BITS);
      expect = a_in + b_in;

      // Don't begin the add operation just yet
      add_en = 0;

      // How many cycles to output valid data?
      val_count = 0;
   end
1: begin
      // This should be the very next clock cycle
      if (val_count !== 1'b0) begin
         $display("\nERROR at time %0t:", $time);
         $display("Valid is not held after reset\n");

         // Use $stop for debugging
         $stop;
      end

      // Test outputs
      if ((sum_out === `BITS'h00000000) &&
           (overflow === 1'b0))
         $display ("Reset is working");
      else begin
         $display("\nERROR at time %0t:", $time);
         $display("Reset is not working");
         $display("    sum_out = %h", sum_out);
         $display("    overflow = %b\n", overflow);
```

```
            // Use $stop for debugging
            $stop;
        end
        // Test valid output signal
        if (valid === 1'b1)
            $display ("Valid signal is working");
        else begin
            $display("\nERROR at time %0t:", $time);
            $display("Valid signal is not working");
            $display("    valid = %b\n", valid);

            // Use $stop for debugging
            $stop;
        end

        // Begin the add operation
        add_en = 1;

        // How many cycles to output valid data?
        val_count = 0;
    end
`TEST_NUM+1: begin
    // Check the result for correctness
    if ({overflow, sum_out} !== expect) begin
        $display("\nERROR at time %0t:", $time);
        $display("Adder is not working");
        $display("    a_in = %h", a_in);
        $display("    b_in = %h", b_in);
        $display("    expected result = %h",
                    expect);
        $display("    actual output = %h\n",
                    {overflow, sum_out});

        // Use $stop for debugging
        $stop;
    end

    // Create random inputs
    // between 0 and all 1s
    a_in = {$random} % (`EBITS'h1 << `BITS);
    b_in = {$random} % (`EBITS'h1 << `BITS);

    // Do not change the expected value
    // since we will  be disabling addition
    add_en = 0;
```

```verilog
            // How many cycles to output valid data?
            val_count = 0;
        end
    `TEST_NUM+2: begin
            // This should be the very next clock cycle
            if (val_count !== 0) begin
                $display("\nERROR at time %0t:", $time);
                $display("Valid is not held\n");

                // Use $stop for debugging
                $stop;
            end
            // Check the result for correctness
            if ({overflow, sum_out} !== expect) begin
                $display("\nERROR at time %0t:", $time);
                $display("Adder is not working");
                $display("    a_in = %h", a_in);
                $display("    b_in = %h", b_in);
                $display("    expected result = %h", expect);
                $display("    actual output = %h\n",
                            {overflow, sum_out});

                // Use $stop for debugging
                $stop;
            end

            $display("\nSimulation complete - no errors\n");
            $finish;
        end
    default: begin
            // Check the result for correctness
            if ({overflow, sum_out} !== expect) begin
                $display("\nERROR at time %0t:", $time);
                $display("Adder is not working");
                $display("    a_in = %h", a_in);
                $display("    b_in = %h", b_in);
                $display("    expected result = %h",
                            expect);
                $display("    actual output = %h\n",
                            {overflow, sum_out});

                // Use $stop for debugging
                $stop;
            end
```

```
                // Create random inputs
                // between 0 and all 1s
                a_in = {$random} % (`EBITS'h1 << `BITS);
                b_in = {$random} % (`EBITS'h1 << `BITS);
                expect = a_in + b_in;

                // Begin the add operation
                add_en = 1;

                // How many cycles to output valid data?
                val_count = 0;
            end
        endcase

        // Count the valid cycles
        cycle_count = cycle_count + 1;
    end
    else begin
        // Keep track of how many cycles to output valid data
        val_count = val_count + 1;

        if (val_count > 11) begin
            $display("\nERROR at time %0t:", $time);
            $display("Too many cycles for valid data\n");

            // Use $stop for debugging
            $stop;
        end
    end
end
endmodule     // add_sim
```

P A R T 3

STATE MACHINES

*S*ince state machines are an essential part of any sequential design, I have decided to devote an entire section to them. Any sequential controller is a state machine. Specifically, in terms of hardware design, a state machine consists of delay elements, combinatorial logic, and feedback from the outputs of the delay elements, through combinatorial logic, back to the inputs of the delay elements. Typically, the delay elements are clocked elements. In other words, the delay is a predictable one, independent of hardware parameters, based on a regular, periodic clock signal. More specifically, the clocked elements are usually D flip-flops. These are the state machines that are easily synthesized and optimized using available synthesis software.

Although the following chapters give advice and information about designing and optimizing state machines, it is important that the synthesis tools that you use can also optimize your state machines. Since state machines control the operation of your design, and are also usually the limiting factor in the clock speed of your design, efficient optimization techniques will produce a faster smaller design, which is always a good thing.

The type of state machines discussed in the following chapters are Moore, Mealy, and One-Hot. Moore and Mealy state machines are two different ways of defining the outputs of the state machine. With a Moore state machine, the outputs are determined by the current state and the current inputs. From a hardware point of view, this means that the outputs consist of taking the state flip-flop outputs and putting them through combinatorial logic to obtain the state machine outputs. With a Mealy state machine, the outputs of the current state are determined by the previous state and the previous inputs. In other words, the outputs are all registered, coming directly from the outputs of flip-flops. This is a definite advantage with respect to hardware design, since the outputs will have the smallest clock-to-output delay, thus creating the least timing constraint on the downstream logic that these outputs affect.

In the past, state machines were optimized to produce the smallest amount of logic. With the advent of Field Programmable Gate Arrays (FPGAs), with fixed architectures, normal state machine optimization may actually increase the utilization of on-chip resources using normal optimization techniques. One-hot state machines are encoded in a way that optimizes them for FPGA use. Both Mealy and Moore state machines can be normally encoded or one-hot encoded. All of these techniques are described in detail in the following chapters.

The Moore State Machine

A Moore state machine is one in which the outputs are determined by the current state of the device. Each state in the state machine corresponds to a specific value for each output. This type of state machine is in contrast to the Mealy state machine, in which each value of an output is associated with a state transition. In other words, in a Moore machine the value of each output depends on the current state and the current values of the inputs. Although either method, Moore or Mealy, can be used to describe any state machine, each has its own advantages. Essentially, the specific requirements of your state machine will determine which method is optimal for your design.

Figure 10-1 shows a very simple state machine for reading and writing a memory device. The state machine begins in the IDLE state after a reset. It then waits in IDLE until it receives a read or a write signal (*rd* or *wr*). If it receives a write signal, it goes to the WRITE state, asserts the write enable (*write_en*) and the acknowledge signal (*ack*) for one cycle. It then returns to the IDLE state, having completed the write operation in a single cycle. If it instead receives a read signal while in the IDLE state, it then goes to the READ1 state and as-

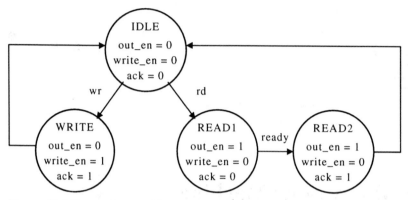

Figure 10-1 Moore state machine.

serts the output enable (out_en). It waits for the ready signal (ready) from the memory device, signaling that valid data is being output. It then goes to the READ2 state, keeping the output enable asserted, and giving an acknowledge, before returning to the IDLE state. Note that for any condition other than ones explicitly shown, it is implied that the state machine remains in the current state. For example, in the READ1 state, if the *ready* signal is not asserted, the machine stays in the READ1 state.

10.1 BEHAVIORAL CODE

```
/*************************************************************/
// MODULE:       Moore state machine
//
// FILE NAME:    moore_beh.v
// VERSION:      1.1
// DATE:         January 1, 2003
// AUTHOR:       Bob Zeidman, Zeidman Consulting
//
// CODE TYPE:    Behavioral Level
//
// DESCRIPTION:  This module shows a state machine
// implementation.
//
// NOTES: This is a Moore model of a state machine, since
// the outputs are defined by the current state.
/*************************************************************/
```

```
// DEFINES
`define DEL   1       // Clock-to-output delay. Zero
                      // time delays can be confusing
                      // and sometimes cause problems.
// TOP MODULE
module state_machine(
      clock,
      reset_n,
      wr,
      rd,
      ready,
      out_en,
      write_en,
      ack);

// PARAMETERS
parameter[1:0]               // State machine states
    IDLE   = 0,
    WRITE  = 1,
    READ1  = 2,
    READ2  = 3;

// INPUTS
input      clock;     // State machine clock
input      reset_n;   // Active low, synchronous reset
input      wr;        // Write command from processor
input      rd;        // Read command from processor
input      ready;     // Ready signal from memory device

// OUTPUTS
output     out_en;    // Output enable to memory
output     write_en;  // Write enable to memory
output     ack;       // Acknowledge signal to processor

// INOUTS

// SIGNAL DECLARATIONS
wire          clock;
wire          reset_n;
wire          wr;
wire          rd;
wire          ready;
reg           out_en;
reg           write_en;
reg           ack;

reg  [1:0]       mem_state;       // Memory state machine
```

```
// ASSIGN STATEMENTS

// MAIN CODE

// Asynchronous reset
always @(posedge reset_n or negedge reset_n) begin
    if (~reset_n)
        #`DEL assign mem_state = IDLE;
    else
        #`DEL deassign mem_state;
end
// Look at the rising edge of clock for state transitions
always @(posedge clock) begin : fsm
    case (mem_state)
        IDLE:  begin
            if (wr == 1'b1)
                mem_state <= #`DEL WRITE;
            else if (rd == 1'b1)
                mem_state <= #`DEL READ1;
        end
        WRITE: begin
            mem_state <= #`DEL IDLE;
        end
        READ1: begin
            if (ready == 1'b1)
                mem_state <= #`DEL READ2;
        end
        READ2: begin
            mem_state <= #`DEL IDLE;
        end
    endcase
end        // fsm

// Look at changes in the state to determine outputs
always @(mem_state) begin : outputs

    // Default output values
    out_en = 1'b0;
    write_en = 1'b0;
    ack = 1'b0;

    case (mem_state)
        WRITE: begin
            write_en = 1'b1;
            ack = 1'b1;
        end
```

```
    READ1: begin
        out_en = 1'b1;
    end
    READ2: begin
        out_en = 1'b1;
        ack = 1'b1;
    end
  endcase
end       // outputs
endmodule     // state_machine
```


10.2 RTL CODE

The following Verilog code shows one RTL implementation of the state machine. In this code, the state machine is represented by the variable, mem_state. The outputs are then determined by an always block that is executed each time the state machine enters a different state. Note that the directive *parallel_case* is used on the *case* statements to help with the state machine optimization. The *full_case* directive is not actually needed since we define all states. Also notice the *state_machine* directive where mem_state is declared. This must be used by some synthesis software to let it know that state machine optimization should be performed. Some synthesis packages can do this automatically without a directive.

```
/***********************************************************/
// MODULE:      Moore state machine
//
// FILE NAME:   moore_rtl.v
// VERSION:     1.1
// DATE:        January 1, 2003
// AUTHOR:      Bob Zeidman, Zeidman Consulting
//
// CODE TYPE:   Register Transfer Level
//
// DESCRIPTION: This module shows a Moore state machine
// implementation.
//
// NOTES: This is a Moore model of a state machine, since
// the outputs are defined by the current state.
/***********************************************************/

// DEFINES
`define DEL   1     // Clock-to-output delay. Zero
                    // time delays can be confusing
                    // and sometimes cause problems.
```

```
// TOP MODULE
module state_machine(
     clock,
     reset_n,
     wr,
     rd,
     ready,
     out_en,
     write_en,
     ack);

// PARAMETERS
parameter[1:0]                    // State machine states
   IDLE   = 0,
   WRITE  = 1,
   READ1  = 2,
   READ2  = 3;

// INPUTS
input     clock;     // State machine clock
input     reset_n;   // Active low, synchronous reset
input     wr;        // Write command from processor
input     rd;        // Read command from processor
input     ready;     // Ready signal from memory device

// OUTPUTS
output    out_en;    // Output enable to memory
output    write_en;  // Write enable to memory
output    ack;       // Acknowledge signal to processor

// INOUTS

// SIGNAL DECLARATIONS
wire         clock;
wire         reset_n;
wire         wr;
wire         rd;
wire         ready;
reg          out_en;
reg          write_en;
reg          ack;

reg  [1:0]     mem_state;          // synthesis state_machine

// ASSIGN STATEMENTS

// MAIN CODE
```

```
// Look at the rising edge of clock for state transitions
always @(posedge clock or negedge reset_n) begin : fsm
   if (~reset_n)
      mem_state <= #`DEL IDLE;
   else begin

                           // use parallel_case directive
                           // to show that all states are
                           // mutually exclusive

      case (mem_state)     // synthesis parallel_case
         IDLE:  begin
            if (wr == 1'b1)
               mem_state <= #`DEL WRITE;
            else if (rd == 1'b1)
               mem_state <= #`DEL READ1;
         end
         WRITE: begin
            mem_state <= #`DEL IDLE;
         end
         READ1: begin
            if (ready == 1'b1)
               mem_state <= #`DEL READ2;
         end
         READ2: begin
            mem_state <= #`DEL IDLE;
         end
      endcase
   end
end        // fsm


// Look at changes in the state to determine outputs
always @(mem_state) begin : outputs

                           // use parallel_case directive
                           // to show that all states are
                           // mutually exclusive
                           // use full_case directive to
                           // show that any undefined
                           // states are don't cares

   case (mem_state)        // synthesis parallel_case full_case
      IDLE:     begin
         out_en = 1'b0;
         write_en = 1'b0;
         ack = 1'b0;
```

```
        end
     WRITE: begin
         out_en = 1'b0;
         write_en = 1'b1;
         ack = 1'b1;
     end
     READ1: begin
         out_en = 1'b1;
         write_en = 1'b0;
         ack = 1'b0;
     end
     READ2: begin
         out_en = 1'b1;
         write_en = 1'b0;
         ack = 1'b1;
     end
  endcase
end       // outputs
endmodule    // state_machine
```

The following Verilog code is an alternative way of implementing the same state machine. In this code, the outputs of each state determine that state, because each state has a unique set of outputs. This is an optimization to which the Moore state machine lends itself. In this way, no flip-flops or extra logic are allocated to state variables, which results in more compact logic. Of course, it will not usually be the case that each state has a distinct, non-intersecting set of outputs. In cases where different states have identical output values, state variables will need to be included to differentiate between them.

```
/**********************************************************/
// MODULE:      Moore state machine
//
// FILE NAME:   moore2_rtl.v
// VERSION:     1.1
// DATE:        January 1, 2003
// AUTHOR:      Bob Zeidman, Zeidman Consulting
//
// CODE TYPE:   Register Transfer Level
//
// DESCRIPTION: This module shows a state machine
// implementation.
//
// NOTES: This is a Moore model of a state machine, since
// the outputs are defined by the current state. This
// implementation uses the outputs to represent the states,
// saving logic.
/**********************************************************/
```

```
// DEFINES
`define DEL   1      // Clock-to-output delay. Zero
                     // time delays can be confusing
                     // and sometimes cause problems.

`define MEM_STATE {out_en, write_en, ack}

// TOP MODULE
module state_machine(
      clock,
      reset_n,
      wr,
      rd,
      ready,
      out_en,
      write_en,
      ack);

// PARAMETERS
parameter[2:0]           // State machine states
    IDLE   = 0,
    WRITE  = 3,
    READ1  = 4,
    READ2  = 5;

// INPUTS
input     clock;     // State machine clock
input     reset_n;   // Active low, synchronous reset
input     wr;        // Write command from processor
input     rd;        // Read command from processor
input     ready;     // Ready signal from memory device

// OUTPUTS
output    out_en;    // Output enable to memory
output    write_en;  // Write enable to memory
output    ack;       // Acknowledge signal to processor

// INOUTS

// SIGNAL DECLARATIONS
wire      clock;
wire      reset_n;
wire      wr;
wire      rd;
wire      ready;
reg       out_en;
reg       write_en;
```

```
reg        ack;

// ASSIGN STATEMENTS

// MAIN CODE

// Look at the rising edge of clock for state transitions
always @(posedge clock or negedge reset_n) begin : fsm
    if (~reset_n)
        `MEM_STATE <= #`DEL IDLE;
    else begin

                            // use parallel_case directive
                            // to show that all states are
                            // mutually exclusive

        case (`MEM_STATE)    // synthesis parallel_case
            IDLE: begin
                if (wr == 1'b1)
                    `MEM_STATE <= #`DEL WRITE;
                else if (rd == 1'b1)
                    `MEM_STATE <= #`DEL READ1;
            end
            WRITE: begin
                `MEM_STATE <= #`DEL IDLE;
            end
            READ1: begin
                if (ready == 1'b1)
                    `MEM_STATE <= #`DEL READ2;
            end
            READ2: begin
                `MEM_STATE <= #`DEL IDLE;
            end
            default: begin   // Since not all states are defined,
                             // include a default state to optimize
                             // synthesis
                `MEM_STATE <= 3'bxxx;
            end
        endcase
    end
end        // fsm
endmodule     // state_machine
```

10.3 SIMULATION CODE

The state machine simulation code uses an extremely simple, one-location memory. The simulation instantiates the memory controller state machine and performs a number of writes and reads to test that the memory is being written and read by the state machine. First, it asserts the reset signal and checks that outputs are reset to the correct values. The simulation then writes a data value to memory, reads it back, and checks that the correct value was read. It reads the memory a second time and checks the value to make sure that the memory did not change after the previous read. This tests back-to-back read operations. Finally, the simulation writes a value to memory and immediately writes a new value to memory. It then reads a value from memory and checks that it was the second value that was read back. This tests back-to-back write operations. Note that tasks are used for read and write operations. Because a number of these operations are performed, this greatly reduced the amount of code that needed to be written.

```
/**********************************************************/
// MODULE:          State machine simulation
//
// FILE NAME:       state_sim.v
// VERSION:         2.1
// DATE:            January 1, 2003
// AUTHOR:          Bob Zeidman, Zeidman Consulting
//
// CODE TYPE:       Simulation
//
// DESCRIPTION:     This module provides stimuli for simulating
// a memory controller state machine. It uses a simple,
// one-location memory and performs a number of writes and
// reads, including back-to-back reads and writes. It also
// checks the reset function.
//
// CHANGES: Version 2.0:
//          1. The minimum memory read access is now three
//             cycles. The state machine can't handle a
//             faster memory than that.
//          2. The simulation waits longer for the ack
//             before complaining that the memory access
//             took too long.
//          3. The simulation performs several cycles of
//             the sequence of reads and writes for better
//             verification coverage.
//
//
/**********************************************************/
```

```verilog
// DEFINES
`define DEL 1            // Clock-to-output delay. Zero
                         // time delays can be confusing
                         // and sometimes cause problems.
`define DWIDTH 8         // Width of data
`define TEST_COUNT 15    // Number of tests to perform

// TOP MODULE
module state_sim();

// PARAMETERS

// INPUTS

// OUTPUTS

// INOUTS

// SIGNAL DECLARATIONS
reg                      clock;
reg                      reset_n;
reg                      write;
reg                      read;
wire                     ready;
wire                     out_en;
wire                     write_en;
wire                     ack;
reg    [`DWIDTH-1:0]     data_out;
wire   [`DWIDTH-1:0]     data;

reg     drd_ready;       // Delayed read ready output
integer cyc_count;       // Count cycles to determine
                         // if the access has taken
                         // too long
integer tcnt;            // count the number of tests

// ASSIGN STATEMENTS
assign #`DEL data = read ? `DWIDTH'hzz : data_out;

// MAIN CODE

// Instantiate the state machine
state_machine machine1(
     .clock(clock),
     .reset_n(reset_n),
```

```verilog
        .wr(write),
        .rd(read),
        .ready(ready),
        .out_en(out_en),
        .write_en(write_en),
        .ack(ack));

// Instantiate a memory
memory memory1(
        .clock(clock),
        .wr(write),
        .rd(read),
        .data(data),
        .ready(ready));

// Generate the clock
always #100 clock = ~clock;

initial begin
   // Initialize inputs
   clock = 0;
   reset_n = 1;
   write = 0;
   read = 0;

   // Test the asynchronous reset
   reset_n = 0;        // Assert the reset signal

   // Wait for the outputs to change asynchronously
   #`DEL
   #`DEL
   // Test outputs
   if ((out_en === 1'b0) && (write_en === 1'b0) &&
      (ack === 1'b0))
      $display ("Reset is working");
   else begin
      $display("\nERROR at time %0t:", $time);
      $display("Reset is not working");
      $display("    out_en = %b", out_en);
      $display("    write_en = %b", write_en);
      $display("    ack = %b\n", ack);

      // Use $stop for debugging
      $stop;
   end
```

```verilog
    // Deassert the reset signal
    reset_n = 1;

    // Perform the sequence multiple times for better
    // verification coverage
    for (tcnt = 0; tcnt < `TEST_COUNT; tcnt = tcnt + 1) begin
        // Write to memory
        writemem(`DWIDTH'h5A);

        // Read from memory
        readmem(`DWIDTH'h5A);

        // Read from memory
        readmem(`DWIDTH'h5A);

        // Write to memory
        writemem(`DWIDTH'h00);

        // Write to memory
        writemem(`DWIDTH'hFF);

        // Read from memory
        readmem(`DWIDTH'hFF);
    end

    $display("\nSimulation complete - no errors\n");
    $finish;
end

always @(posedge clock) begin
    // Create delayed read ready
    drd_ready <= #`DEL (ready && read);

    // Check whether an access is taking too long
    if (cyc_count > 8) begin
        $display("\nERROR at time %0t:", $time);
        $display("Read access took too long\n");

        // Use $stop for debugging
        $stop;
    end

    // Check the ack output
    if (write_en) begin
        if (~ack) begin
            $display("\nERROR at time %0t:", $time);
```

```
            $display("Write access - ack is not asserted\n");

            // Use $stop for debugging
            $stop;
        end
    end
    else if (drd_ready) begin
        if (~ack) begin
            $display("\nERROR at time %0t:", $time);
            $display("Read access - ack is not asserted\n");

            // Use $stop for debugging
            $stop;
        end
    end
    else if (ack) begin
        $display("\nERROR at time %0t:", $time);
        $display("No access - ack is asserted\n");

        // Use $stop for debugging
        $stop;
    end
end

// TASKS
// Write data to memory
task writemem;

// INPUTS
input [`DWIDTH-1:0]  write_data;    // Data to write to mem-
ory

// OUTPUTS

// INOUTS

// TASK CODE
begin
    // Wait for the rising clock edge
    @(posedge clock);
    read <= #`DEL 0;        // Deassert the read controls
    write <= #`DEL 1;       // Assert the write control
    data_out <= write_data;

    // Wait for one cycle
    @(posedge clock);
```

```
end
endtask     // writemem

// Read data from memory and check its value
task readmem;

// INPUTS
input [`DWIDTH-1:0]   expected;    // Expected read data

// OUTPUTS

// INOUTS

// TASK CODE
begin
    cyc_count = 0;              // Initialize the cycle count

    // Wait for the rising clock edge
    @(posedge clock);
    read <= #`DEL 1;           // Assert the read control
    write <= #`DEL 0;          // Deassert the write control
    data_out = `DWIDTH'hxx;    // Put out undefined data

    // Wait for the ready signal
    @(posedge clock);
    while (~ready) begin
        // Increment the cycle count
        cyc_count = cyc_count + 1;

        @(posedge clock);
    end

    // Did we find the expected data?
    if (ready === 1) begin
        if (data === expected) begin
            $display ("Memory is working");
        end
        else begin
            $display("\nERROR at time %0t:", $time);
            $display("Memory is not working");
            $display("    data written = %h", expected);
            $display("    data read = %h\n", data);

            // Use $stop for debugging
            $stop;
        end
```

```
      end

end
endtask      // readmem

endmodule           // state_sim

// SUBMODULE
// One byte memory for simulation
module memory(
      clock,
      data,
      wr,
      rd,
      ready);

// INPUTS
input       clock;      // Clock input
input       wr;         // Write input
input       rd;         // Read input

// OUTPUTS
output      ready;      // Is the memory ready?

// INOUTS
inout [`DWIDTH-1:0]  data;     // Data lines

// SIGNAL DECLARATIONS
wire                 clock;
wire                 wr;
wire                 rd;
reg                  ready;
wire [`DWIDTH-1:0]   data;
reg  [`DWIDTH-1:0]   mem;       // Stored data

reg  [2:0]           count;     // Counter to assert ready

// PARAMETERS

// ASSIGN STATEMENTS
assign #`DEL data = rd ? mem : `DWIDTH'bzz;

// MAIN CODE
initial begin
   count = $random(0);   // Initialize random number generator
   count = 0;
```

```
end

always @(posedge clock) begin
    ready <= #`DEL 1'b0;
    if ((wr === 1) && ~ready) begin
        mem <= #`DEL data;
        ready <= #`DEL 1'b1;
    end
    else if (rd === 1) begin
        if (count === 0) begin
            count = {$random} % 6 + 2;
        end
        else begin
            count = count - 1;
            if (count === 0) begin
                ready <= #`DEL 1'b1;
            end
        end
    end
end
endmodule            // memory
```

The Mealy State Machine

A Mealy state machine is one in which the outputs change on the transitions of the device. In other words, the outputs for the next state are determined by the current state and the current inputs. This type of state machine is in contrast to the Moore state machine, in which each value of an output is associated with a specific current state. In a Mealy machine, the outputs during a particular state can be different at different times, depending on how the machines entered the current state While either method, Mealy or Moore, can be used to describe any state machine, each has its own advantages. Essentially the specific requirements of the state machine will determine which method is optimal for your design.

Figure 11-1 shows a very simple state machine for reading and writing a memory device. The state machine begins in the IDLE state after a reset. It then waits in IDLE until it receives a read or a write signal (*rd* or *wr*). If it receives a write signal, it goes to the WRITE state, asserts the write enable (*write_en*) and the acknowledge signal (*ack*) for one cycle. It then returns to the IDLE state, having completed the write operation in a single cycle. If it receives a read signal in the IDLE state, it goes to the READ1 state and asserts

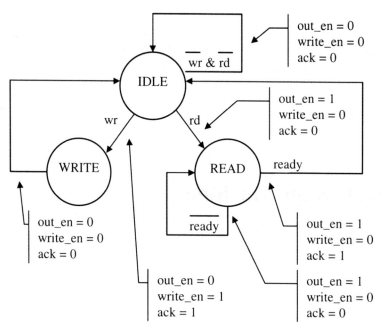

Figure 11-1 Mealy state machine.

the output enable (out_en). It waits for the ready signal (ready) from the memory device, signaling that valid data is being output. It then goes to the READ2 state, keeping the output enable asserted, and giving an acknowledge, before returning to the IDLE state. Note that each state transition needs to be shown, even those default transitions where the machine remains in the same state, because outputs change for each transition.

11.1 BEHAVIORAL CODE

```
/******************************************************************/
// MODULE:       Mealy state machine
//
// FILE NAME:    mealy_beh.v
// VERSION:      1.1
// DATE:         January 1, 2003
// AUTHOR:       Bob Zeidman, Zeidman Consulting
//
// CODE TYPE:    Behavioral Level
//
// DESCRIPTION:  This module shows a state machine
// implementation.
//
```

```
// NOTES: This is a Mealy model of a state machine, since
// the outputs are defined by the state transitions.
/************************************************************/

// DEFINES
`define DEL   1      // Clock-to-output delay. Zero
                     // time delays can be confusing
                     // and sometimes cause problems.

// TOP MODULE
module state_machine(
     clock,
     reset_n,
     wr,
     rd,
     ready,
     out_en,
     write_en,
     ack);

// PARAMETERS
parameter[1:0]            // State machine states
    IDLE  = 0,
    WRITE = 1,
    READ  = 2;

// INPUTS
input         clock;    // State machine clock
input         reset_n;  // Active low, synchronous reset
input         wr;       // Write command from processor
input         rd;       // Read command from processor
input         ready;    // Ready signal from memory device

// OUTPUTS
output        out_en;   // Output enable to memory
output        write_en; // Write enable to memory
output        ack;      // Acknowledge signal to processor

// INOUTS

// SIGNAL DECLARATIONS
wire          clock;
wire          reset_n;
wire          wr;
wire          rd;
wire          ready;
reg   [1:0]   mem_state;       // Memory state machine
```

```
reg            out_en;
reg            write_en;
reg            ack;

reg   [1:0]       mem_state;         .// Memory state machine

// ASSIGN STATEMENTS

// MAIN CODE

// Asynchronous reset
always @(posedge reset_n or negedge reset_n) begin
    if (~reset_n) begin
        #`DEL;
        assign mem_state = IDLE;
        assign out_en = 1'b0;
        assign write_en = 1'b0;
        assign ack = 1'b0;
    end
    else begin
        #`DEL;
        deassign mem_state;
        deassign out_en;
        deassign write_en;
        deassign ack;
    end
end

// Look at the rising edge of clock for state transitions
always @(posedge clock) begin : fsm
    case (mem_state)
        IDLE:  begin
            // Do not do anything if we are
            // acknowledging a previous access
            if (~ack && (wr == 1'b1)) begin
                mem_state <= #`DEL WRITE;
                out_en <= #`DEL 1'b0;
                write_en <= #`DEL 1'b1;
                ack <= #`DEL 1'b1;
            end
            else if (~ack && (rd == 1'b1)) begin
                mem_state <= #`DEL READ;
                out_en <= #`DEL 1'b1;
                write_en <= #`DEL 1'b0;
                ack <= #`DEL 1'b0;
            end
            else begin
                out_en <= #`DEL 1'b0;
```

```
                    write_en <= #`DEL 1'b0;
                    ack <= #`DEL 1'b0;
                end
            end
            WRITE: begin
                mem_state <= #`DEL IDLE;
                out_en <= #`DEL 1'b0;
                write_en <= #`DEL 1'b0;
                ack <= #`DEL 1'b0;
            end
            READ:  begin
                if (ready == 1'b1) begin
                    mem_state <= #`DEL IDLE;
                    out_en <= #`DEL 1'b1;
                    write_en <= #`DEL 1'b0;
                    ack <= #`DEL 1'b1;
                end
                else begin
                    out_en <= #`DEL 1'b1;
                    write_en <= #`DEL 1'b0;
                    ack <= #`DEL 1'b0;
                end
            end
        endcase
end    // fsm
endmodule    // state_machine
```

11.2 RTL Code

Note that the directive *parallel_case* is used on the *case* statement to help with the state machine optimization. The *full_case* directive is not actually needed since we define all states.

```
/*********************************************************/
// MODULE:       Mealy state machine
//
// FILE NAME:    mealy_rtl.v
// VERSION:      1.1
// DATE:         January 1, 2003
// AUTHOR:       Bob Zeidman, Zeidman Consulting
//
// CODE TYPE:    Register Transfer Level
//
// DESCRIPTION:  This module shows a state machine
// implementation.
```

```
//
// NOTES: This is a Mealy model of a state machine, since
// the outputs are defined by the current state.
/*********************************************************/

// DEFINES
`define DEL   1      // Clock-to-output delay. Zero
                     // time delays can be confusing
                     // and sometimes cause problems.

// TOP MODULE
module state_machine(
      clock,
      reset_n,
      wr,
      rd,
      ready,
      out_en,
      write_en,
      ack);

// PARAMETERS
parameter[1:0]                 // State machine states
    IDLE   = 0,
    WRITE  = 1,
    READ   = 2;

// INPUTS
input          clock;      // State machine clock
input          reset_n;    // Active low, synchronous reset
input          wr;         // Write command from processor
input          rd;         // Read command from processor
input          ready;      // Ready signal from memory device

// OUTPUTS
output         out_en;     // Output enable to memory
output         write_en;   // Write enable to memory
output         ack;        // Acknowledge signal to processor

// INOUTS

// SIGNAL DECLARATIONS
wire           clock;
wire           reset_n;
wire           wr;
wire           rd;
wire           ready;
```

```
reg             out_en;
reg             write_en;
reg             ack;

reg  [1:0]      mem_state;      // synthesis state_machine

// ASSIGN STATEMENTS

// MAIN CODE

// Look at the rising edge of clock for state transitions
always @(posedge clock or negedge reset_n) begin : fsm
   if (~reset_n) begin
      mem_state <= #`DEL IDLE;
      out_en <= #`DEL 1'b0;
      write_en <= #`DEL 1'b0;
      ack <= #`DEL 1'b0;
   end
   else begin
                        // use parallel_case directive
                        // to show that all states are
                        // mutually exclusive

      case (mem_state)    // synthesis parallel_case
         IDLE: begin
            // Do not do anything if we are
            // acknowledging a previous access
            if (~ack && (wr == 1'b1)) begin
               mem_state <= #`DEL WRITE;
               out_en <= #`DEL 1'b0;
               write_en <= #`DEL 1'b1;
               ack <= #`DEL 1'b1;
            end
            else if (~ack && (rd == 1'b1)) begin
               mem_state <= #`DEL READ;
               out_en <= #`DEL 1'b1;
               write_en <= #`DEL 1'b0;
               ack <= #`DEL 1'b0;
            end
            else begin
               out_en <= #`DEL 1'b0;
               write_en <= #`DEL 1'b0;
               ack <= #`DEL 1'b0;
            end
         end
         WRITE: begin
            mem_state <= #`DEL IDLE;
            out_en <= #`DEL 1'b0;
```

```
                    write_en <= #`DEL 1'b0;
                    ack <= #`DEL 1'b0;
                end
                READ: begin
                    if (ready == 1'b1) begin
                        mem_state <= #`DEL IDLE;
                        out_en <= #`DEL 1'b1;
                        write_en <= #`DEL 1'b0;
                        ack <= #`DEL 1'b1;
                    end
                    else begin
                        out_en <= #`DEL 1'b1;
                        write_en <= #`DEL 1'b0;
                        ack <= #`DEL 1'b0;
                    end
                end
                default: begin   // Since not all states are defined,
                                 // include a default state to
                                 // optimize synthesis
                    mem_state <= #`DEL 2'bxx;
                    out_en <= #`DEL 1'bx;
                    write_en <= #`DEL 1'bx;
                    ack <= #`DEL 1'bx;
                end
            endcase
        end
end     // fsm
endmodule    // state_machine
```

11.3 SIMULATION CODE

The code used to simulate this module is the same simulation code used in Chapter 10.

The One-Hot State Machine for FPGAs

For Field Programmable Gate Arrays (FPGAs), the normal method of designing state machines is not optimal. This is because FPGAs are made up of logic blocks, unlike Application Specific Integrated Circuits (ASICs), which can be designed using simple logic gates. Each logic block in an FPGA has one or more flip-flops plus some limited combinatorial logic, making for an abundance of flip-flops. For creating large combinatorial logic terms, however, many logic blocks are often involved, which requires connecting these blocks through slow interconnect. If we look at the RTL Verilog code for the simple memory controller state machine in Chapter 10, the resulting gate level implementation will probably look like Figure 12-1 after synthesis. Notice that it uses few flip-flops and much combinatorial logic. This is good for ASICs, bad for FPGAs.

The better method of designing state machines for FPGAs is known as one-hot encoding, seen in Figure 12-2. Using this method, each state is represented by a single flip-flop, rather than encoded from several flip-flop outputs. This greatly reduces the combina-

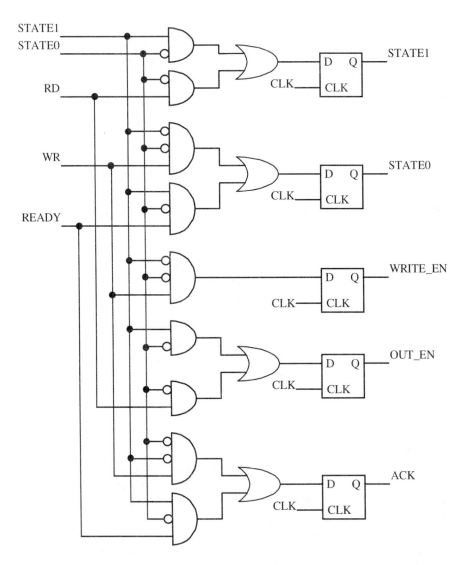

Figure 12-1 Normally encoded state machine after synthesis.

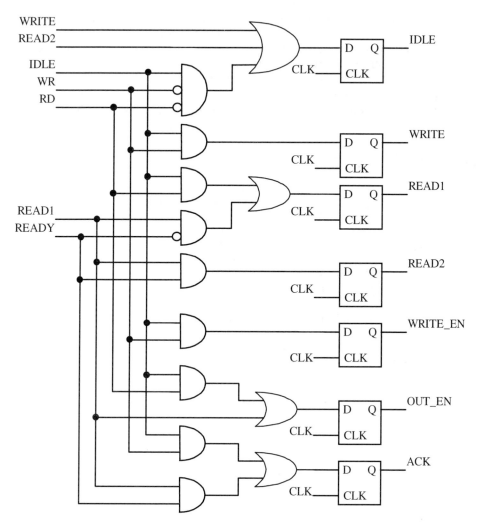

Figure 12-2 One-hot encoded state machine after synthesis.

torial logic, since only one bit needs to be checked to see if the state machine is in a particular state. Notice that although there are more flip-flops generated using one-hot encoding, there are fewer gates per flip-flop. Also, the gates generated have fewer inputs and are thus easier to fit into the limited combinatorial logic of an FPGA logic block. The more complex the state machine, the greater this effect will be.

It is important to note that each state bit flip-flop needs to be reset when initialized, except for the IDLE state flip-flop that needs to be set so that the state machine begins in the IDLE state.

12.1 RTL CODE

```
/***********************************************************/
// MODULE:         One-Hot state machine
//
// FILE NAME:      onehot_rtl.v
// VERSION:        1.1
// DATE:           January 1, 2003
// AUTHOR:         Bob Zeidman, Zeidman Consulting
//
// CODE TYPE:      Register Transfer Level
//
// DESCRIPTION:  This module shows a state machine
// implemented using one-hot encoding.
//
// NOTES:
/***********************************************************/

// DEFINES
`define DEL   1      // Clock-to-output delay. Zero
                     // time delays can be confusing
                     // and sometimes cause problems.

// TOP MODULE
module state_machine(
      clock,
      reset_n,
      wr,
      rd,
      ready,
      out_en,
      write_en,
      ack);

// PARAMETERS
parameter[3:0]              // State machine states
    IDLE_STATE     = 4'b0001,
```

```
    WRITE_STATE   = 4'b0010,
    READ1_STATE   = 4'b0100,
    READ2_STATE   = 4'b1000;

parameter[3:0]                    // State machine cases
    IDLE_CASE     = 4'bxxx1,
    WRITE_CASE    = 4'bxx1x,
    READ1_CASE    = 4'bx1xx,
    READ2_CASE    = 4'b1xxx;

// INPUTS
input     clock;      // State machine clock
input     reset_n;    // Active low, synchronous reset
input     wr;         // Write command from processor
input     rd;         // Read command from processor
input     ready;      // Ready signal from memory device

// OUTPUTS
output    out_en;     // Output enable to memory
output    write_en;   // Write enable to memory
output    ack;        // Acknowledge signal to processor

// INOUTS

// SIGNAL DECLARATIONS
wire          clock;
wire          reset_n;
wire          wr;
wire          rd;
wire          ready;
reg           out_en;
reg           write_en;
reg           ack;
reg  [3:0]    mem_state;         // synthesis state_machine

// ASSIGN STATEMENTS

// MAIN CODE

// Look at the rising edge of clock for state transitions
always @(posedge clock or negedge reset_n) begin : fsm
   if (~reset_n)
      mem_state <= #`DEL IDLE_STATE;
   else begin
                            // use parallel_case directive to
                            // to show that all states are
                            // mutually exclusive
                            // use full_case directive to
```

```
                              // show that any undefined states
                              // are don't cares
                              // use casex statement to show that
                              // some bit positions are don't cares

        casex (mem_state)    // synthesis parallel_case full_case
            IDLE_CASE:    begin
                if (wr == 1'b1)
                    mem_state <= #`DEL WRITE_STATE;
                else if (rd == 1'b1)
                    mem_state <= #`DEL READ1_STATE;
            end
            WRITE_CASE:  begin
                mem_state <= #`DEL IDLE_STATE;
            end
            READ1_CASE:  begin
                if (ready == 1'b1)
                    mem_state <= #`DEL READ2_STATE;
            end
            READ2_CASE:  begin
                mem_state <= #`DEL IDLE_STATE;
            end
        endcase
    end
end         // fsm

// Look at changes in the state to determine outputs
always @(mem_state) begin : outputs

                              // use parallel_case directive
                              // to show that all states are
                              // mutually exclusive
                              // use full_case directive to
                              // show that any undefined states
                              // are don't cares
                              // use casex statement to show that
                              // some bit positions are don't cares

    casex (mem_state)        // synthesis parallel_case full_case
        IDLE_CASE:    begin
            out_en = 1'b0;
            write_en = 1'b0;
            ack = 1'b0;
        end
        WRITE_CASE:  begin
            out_en = 1'b0;
            write_en = 1'b1;
```

```
          ack = 1'b1;
      end
      READ1_CASE:   begin
          out_en = 1'b1;
          write_en = 1'b0;
          ack = 1'b0;
      end
      READ2_CASE:   begin
          out_en = 1'b1;
          write_en = 1'b0;
          ack = 1'b1;
      end
   endcase
end        // outputs
endmodule    // state_machine
```

12.2 SIMULATION CODE

The code used to simulate this module is the same simulation code used in Chapter 10.

P A R T   **4**

MISCELLANEOUS COMPLEX FUNCTIONS

In the following chapters, I discuss and examine code for some more complex functions than those that have already been described. These special purpose functions find use in many common circuits.

The Linear Feedback Shift Register (LFSR)

The Linear Feedback Shift Register (LFSR) is an extremely useful function that, I believe, would be used even more often if there were more good references showing how to design one. Most textbooks that cover LFSRs, and there aren't many of them that I've been able to find, discuss the mathematical theory of primitive polynomials. Unfortunately, that doesn't help an electrical engineer design one from flip-flops and combinatorial logic. There is one book, *Designus Maximus Unleashed!* by Clive "Max" Maxfield who writes for *EDN magazine*. It's a very fun, off-the-wall, and extremely practical book full of down to earth explanations and examples. I've borrowed some of the information from that book, in order to explain LFSRs, with Max's gracious permission.

In mathematical terms, an LFSR is a representation of a primitive polynomial. In more practical language, it is a method of generating a pseudo-random sequence of numbers. In even simpler language, imagine taking slips of paper with all of the numbers from 1 to some number 2^n-1, putting them in a hat, and having a blindfolded assistant draw them out one by one. The output of an LFSR would look like that sequence of numbers. It's not truly random, because we can predict the entire sequence. However, the numbers are randomly distributed.

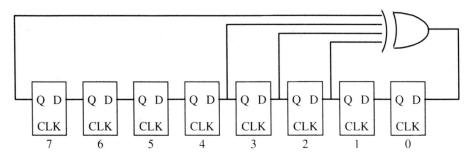

Figure 13-1 An 8-bit linear feedback shift register (LFSR). *Many 2 One*

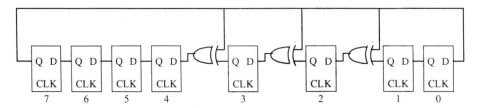

Figure 13-2 An alternate 8-bit linear feedback shift register (LFSR).

One 2 many

The basic functional diagrams for an LFSR are shown in Figures 13-1 and 13-2. These are two basic ways of implementing the LFSR, and both produce a pseudo-random sequence, though not the same sequence. The implementation in Figure 31-1 requires fewer gates, whereas the implementation in Figure 13-2 may have slightly better timing since there is only a two input XOR gate between any two registers. Both LFSRs will output the numbers from 1 to 2n-1 where n is the number of bits (8 in this case). Also, the XOR gates can be replaced by XNOR gates. Again, the sequence will be a random distribution, but it will not be the same sequence.

Notice that both implementations use the same taps. Taps are where we tap into the shift register. In the first example, the taps for the XOR gate are placed at the outputs of bits 7, 3, 2, and 1. In the second example, bit 7 is fed into bit 0, and also combined with the bits 1, 2, and 3. In both cases, the essential taps are bits 7, 3, 2, and 1. Not all combinations of taps will produce the correct sequence, although there are more than one set of good taps. One list of good taps is given in Table 13-1. If a bad set of taps is used, the LFSR will not produce all possible numbers, but only a limited set of them. Also, note that the number 0 will not appear in the sequence. If a 0 is loaded into the LFSR, it will remain 0 for each clock cycle thereafter (when XNOR gates are used, a value of all ones will result in a steady, unchanging state). For this reason, when the LFSR is initialized, it must have some non-zero value (or not all ones if you are using XNORs). If the 0 value is required, it can be done with the simple addition of a NOR gate and either an extra XOR gate or a bigger XOR gate, depending on the LFSR implementation. This logic inserts a 0 value into the sequence, and is shown in Figure 13-3 and Figure 13-4 for the two different implementations. The disadvantage of this extra logic is not only extra hardware, but technically the sequence is no longer random because the 0 value always occurs at a specific place in the sequence. This may or may not be a problem for you, depending on the particular use of your LFSR.

The uses of the LFSR are many. Obviously, their ability to generate random sequences of numbers is very useful for Built-In Self Test (BIST) circuits, where a test circuit is needed to supply inputs to the main circuit in order to determine whether the main circuit is working correctly. Applying a long sequence of random inputs and comparing the outputs to the expected outputs will tell if the circuit is working correctly. Of course it would be ideal to generate meaningful inputs, but to do so would require a very large amount of on-chip circuitry. An LFSR generates a large sequence of inputs with very little hardware. Similarly, the resulting outputs can be combined, using an LFSR, into a "signature" for the hardware being tested. Rather than examining each output for each bit the entire input sequence, only the simple signature needs to be compared to the expected signature once the sequence has finished. Again, the LFSR significantly reduces the amount of extra test hardware needed. If the correct signature is created from the random sequence of inputs, the probability is extremely high that the circuit is working correctly.

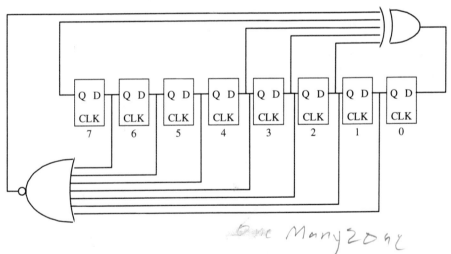

Figure 13-3 8-bit linear feedback shift register (LFSR) with 0 value inserted.

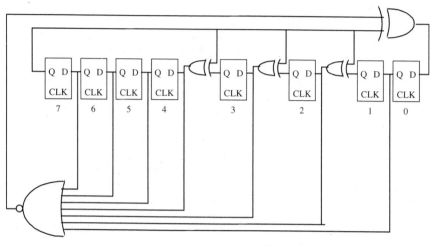

Ou[Zman-]

Figure 13-4 Alternate 8-bit linear feedback shift register (LFSR) with 0 value inserted.

$Data[MAX-1:1] = (data ∧ ‾TAPS) << 1$

Table 13-1   LFSR Taps $D[0] = D[MAX-1] ∧ ~|(data ∧ TAPS)$

Number of Bits (n)	Sequence length (2ⁿ-1)	Tap Locations
2	3	1,0
3	7	2,0
4	15	3,0
5	31	4,1
6	63	5,0
7	127	6,0
8	255	7,3,2,1
9	511	8,3
10	1,023	9,2
11	2,047	10,1
12	4,095	11,5,3,0
13	8,191	12,3,2,0
14	16,383	13,4,2,0
15	32,767	14,0
16	65,535	15,4,2,1
17	131,071	16,2
18	262,143	17,6

Number of Bits (n)	Sequence length (2^n-1)	Tap Locations
19	524,287	18,4,1,0
20	1,048,575	19,2
21	2,097,151	20,1
22	4,194,303	21,0
23	8,388,607	22,4
24	16,777,215	23,3,2,0
25	33,554,431	24,2
26	67,108,863	25,5,1,0
27	134,217,727	26,4,1,0
28	268,435,455	27,2
29	536,870,911	28,1
30	1,073,741,823	29,5,3,0
31	2,147,483,647	30,2
32	4,294,967,295	31,6,5,1

[handwritten annotations in right margin: "24,7 www.maxim-ic.com/appnotes.csm/an_pk/1743 (rest same)" near row 25; "all same as Smith b" near rows 31-32]

LFSRs are also used to generate Cyclic Redundancy Checks (CRCs), a sophisticated form of error checking that is similar, but more reliable than, a checksum. A CRC is used to determine whether data had been stored or transmitted without errors. An example CRC design is given in a later chapter in this book.

Encryption and decryption hardware is also dependent on the LFSR. An example of encryption and decryption hardware is given in another chapter in this book.

13.1 BEHAVIORAL CODE

The following sections show behavioral code for the two different implementations of an LFSR. Note that there are conditionally compiled statements between the `ifdef`, `else`, and `endif` directives. If your design requires that the value of zero be inserted into the sequence of numbers, simply define the add_zero constant. If you do not require the zero value, do nothing and leave the add_zero constant undefined.

13.1.1 Implementation 1

```
/*****************************************************/
// MODULE:        linear feedback shift register (LFSR)
```

```
//
// FILE NAME:    lfsr_beh.v
// VERSION:      1.1
// DATE:         January 1, 2003
// AUTHOR:       Bob Zeidman, Zeidman Consulting
//
// CODE TYPE:    Behavioral Level
//
// DESCRIPTION:  This module defines a linear feedback shift
// register (LFSR) using a single XOR gate.
//                              Many to one
/***********************************************************/

// DEFINES
`define DEL    1       // Clock-to-output delay. Zero
                       // time delays can be confusing
                       // and sometimes cause problems.

                       // These are good tap values for 2 to 32 bits
`define TAP2     2'b11
`define TAP3     3'b101
`define TAP4     4'b1001
`define TAP5     5'b10010
`define TAP6     6'b100001
`define TAP7     7'b1000001
`define TAP8     8'b10001110
`define TAP9     9'b100001000
`define TAP10   10'b1000000100
`define TAP11   11'b10000000010
`define TAP12   12'b100000101001            Same as sixth
`define TAP13   13'b1000000001101
`define TAP14   14'b10000000010101
`define TAP15   15'b100000000000001
`define TAP16   16'b1000000000010110
`define TAP17   17'b10000000000000100
`define TAP18   18'b100000000001000000
`define TAP19   19'b1000000000000010011
`define TAP20   20'b10000000000000000100
`define TAP21   21'b100000000000000000010
`define TAP22   22'b1000000000000000000001
`define TAP23   23'b10000000000000000010000
`define TAP24   24'b100000000000000000001101
`define TAP25   25'b1000000000000000000000100
`define TAP26   26'b10000000000000000000100011
`define TAP27   27'b100000000000000000000010011
`define TAP28   28'b1000000000000000000000000100
`define TAP29   29'b10000000000000000000000000010
`define TAP30   30'b100000000000000000000000101001
```

```
`define TAP31    31'b1000000000000000000000000000100
`define TAP32    32'b10000000000000000000000001100010

`define BITS 8              // Number of bits in the LFSR
`define TAPS `TAP8          // This must be the taps for the
                           // number of bits specified above
`define INIT 1              // This can be any non-zero value
                           // for initialization of the LFSR

// TOP MODULE
module LFSR(
     clk,
     reset,
     data);

// PARAMETERS

// INPUTS
input               clk;      // Clock
input               reset;    // Reset

// OUTPUTS
output [`BITS-1:0]  data;     // LFSR data

// INOUTS

// SIGNAL DECLARATIONS
wire                clk;
wire                reset;
reg [`BITS-1:0]     data;

// ASSIGN STATEMENTS

// MAIN CODE

// Look at the edges of reset
always @(posedge reset or negedge reset) begin
   if (reset)
       #`DEL assign data = `INIT;
   else
       #`DEL deassign data;
end

// Look at the rising edge of clock for state transitions
always @(posedge clk) begin
   // Shift all of the bits left
   data[`BITS-1:1] <= #`DEL data[`BITS-2:0];
```

```
`ifdef ADD_ZERO  // Use this code if data == 0 is required
   // Create the new bit 0
   data[0] <= #`DEL ^(data & `TAPS) ^ ~|data[`BITS-2:0];
`else                // Use this code for a standard LFSR
   // Create the new bit 0
   data[0] <= #`DEL ^(data & `TAPS);
`endif

end
endmodule    // LFSR
```

13.1.2 Implementation 2

```
/*****************************************************************/
// MODULE:      linear feedback shift register (LFSR)
//
// FILE NAME:   lfsr2_beh.v
// VERSION:     1.1
// DATE:        January 1, 2003
// AUTHOR:      Bob Zeidman, Zeidman Consulting
//
// CODE TYPE:   Behavioral Level
//
// DESCRIPTION: This module defines a linear feedback shift
// register (LFSR) using multiple XOR gates.
//
/*****************************************************************/

// DEFINES
`define DEL  1        // Clock-to-output delay. Zero
                      // time delays can be confusing
                      // and sometimes cause problems.

                      // These are good tap values for 2 to 32 bits
`define TAP2    2'b11
`define TAP3    3'b101
`define TAP4    4'b1001
`define TAP5    5'b10010
`define TAP6    6'b100001
`define TAP7    7'b1000001
`define TAP8    8'b10001110
`define TAP9    9'b100001000
`define TAP10   10'b1000000100
`define TAP11   11'b10000000010
```

```
`define TAP12    12'b100000101001
`define TAP13    13'b1000000001101
`define TAP14    14'b10000000010101
`define TAP15    15'b100000000000001
`define TAP16    16'b1000000000010110
`define TAP17    17'b10000000000000100
`define TAP18    18'b100000000001000000
`define TAP19    19'b1000000000000010011
`define TAP20    20'b10000000000000000100
`define TAP21    21'b100000000000000000010
`define TAP22    22'b1000000000000000000001
`define TAP23    23'b10000000000000000010000
`define TAP24    24'b100000000000000000001101
`define TAP25    25'b1000000000000000000000100
`define TAP26    26'b10000000000000000000100011
`define TAP27    27'b100000000000000000000010011
`define TAP28    28'b1000000000000000000000000100
`define TAP29    29'b10000000000000000000000000010  → ?
`define TAP30    30'b100000000000000000000000101001
`define TAP31    31'b1000000000000000000000000000100
`define TAP32    32'b100000000000000000000000001100010

`define BITS 8            // Number of bits in the LFSR
`define TAPS `TAP8        // This must be the taps for the
                          // number of bits specified above
`define INIT 1            // This can be any non-zero value
                          // for initialization of the LFSR

// TOP MODULE
module LFSR(
      clk,
      reset,
      data);

// PARAMETERS

// INPUTS
input              clk;      // Clock
input              reset;    // Reset

// OUTPUTS
output [`BITS-1:0]  data;     // LFSR data

// INOUTS

// SIGNAL DECLARATIONS
wire               clk;
wire               reset;
```

```verilog
reg [`BITS-1:0]      data;

// ASSIGN STATEMENTS

// MAIN CODE

// Look at the edges of reset
always @(posedge reset or negedge reset) begin
    if (reset)
        #`DEL assign data = `INIT;
    else
        #`DEL deassign data;
end

// Look at the rising edge of clock for state transitions
always @(posedge clk) begin
    // XOR each tap bit with the most significant bit
    // and shift left. Also XOR the most significant
    // bit with the incoming bit
    if (data[`BITS-1]) begin
        data <= #`DEL (data ^ `TAPS) << 1;

`ifdef ADD_ZERO // Use this code if data == 0 is required
        data[0] <= #`DEL data[`BITS-1] ^ ~|(data ^ `TAPS);
`else
        data[0] <= #`DEL data[`BITS-1];
`endif

    end
    else begin
        data <= #`DEL data << 1;

`ifdef ADD_ZERO // Use this code if data == 0 is required
        data[0] <= #`DEL data[`BITS-1] ^ ~|data;
`else
        data[0] <= #`DEL data[`BITS-1];
`endif

    end
end
endmodule      // LFSR
```

13.2 RTL CODE

The following sections show RTL code for the two different implementations of an LFSR. Note that there are conditionally compiled statements between the `ifdef`, `else`, and `endif` directives. If your design requires that the value of zero be inserted into the sequence of numbers, simply define the add_zero constant. If you do not require the zero value, do nothing and leave the add_zero constant undefined.

13.2.1 Implementation 1

```
/*********************************************************/
// MODULE:        linear feedback shift register (LFSR)
//
// FILE NAME:     lfsr_rtl.v
// VERSION:       1.1
// DATE:          January 1, 2003
// AUTHOR:        Bob Zeidman, Zeidman Consulting
//
// CODE TYPE:     Register Transfer Level
//
// DESCRIPTION:   This module defines a linear feedback shift
// register (LFSR) using a single XOR gate.
//
/*********************************************************/

// DEFINES
`define DEL   1     // Clock-to-output delay. Zero
                    // time delays can be confusing
                    // and sometimes cause problems.

                    // These are good tap values for 2 to 32 bits
`define TAP2      2'b11
`define TAP3      3'b101
`define TAP4      4'b1001
`define TAP5      5'b10010
`define TAP6      6'b100001
`define TAP7      7'b1000001
`define TAP8      8'b10001110
`define TAP9      9'b100001000
`define TAP10    10'b1000000100
`define TAP11    11'b10000000010
`define TAP12    12'b100000101001
`define TAP13    13'b1000000001101
`define TAP14    14'b10000000010101
```

```verilog
`define TAP15    15'b100000000000001
`define TAP16    16'b1000000000010110
`define TAP17    17'b10000000000000100
`define TAP18    18'b100000000001000000
`define TAP19    19'b1000000000000010011
`define TAP20    20'b10000000000000000100
`define TAP21    21'b100000000000000000010
`define TAP22    22'b1000000000000000000001
`define TAP23    23'b10000000000000000010000
`define TAP24    24'b100000000000000000001101
`define TAP25    25'b1000000000000000000000100
`define TAP26    26'b10000000000000000000100011
`define TAP27    27'b100000000000000000000010011
`define TAP28    28'b1000000000000000000000000100
`define TAP29    29'b10000000000000000000000000010
`define TAP30    30'b100000000000000000000000101001
`define TAP31    31'b1000000000000000000000000000100
`define TAP32    32'b10000000000000000000000001100010

`define BITS 8          // Number of bits in the LFSR
`define TAPS `TAP8       // This must be the taps for the
                         // number of bits specified above
`define INIT 1           // This can be any non-zero value
                         // for initialization of the LFSR

// TOP MODULE
module LFSR(
      clk,
      reset,
      data);

// PARAMETERS

// INPUTS
input                clk;      // Clock
input                reset;    // Reset

// OUTPUTS
output [`BITS-1:0]  data;      // LFSR data

// INOUTS

// SIGNAL DECLARATIONS
wire                clk;
wire                reset;
reg [`BITS-1:0]     data;
```

```
// ASSIGN STATEMENTS

// MAIN CODE

// Look at the rising edge of clock or reset
always @(posedge clk or posedge reset) begin
    if (reset)
        data <= #`DEL `INIT;
    else begin
        // Shift all of the bits left
        data[`BITS-1:1] <= #`DEL data[`BITS-2:0];

`ifdef ADD_ZERO  // Use this code if data == 0 is required
        // Create the new bit 0
        data[0] <= #`DEL ^(data & `TAPS) ^ ~|data[`BITS-2:0];
`else                    // Use this code for a standard LFSR
        // Create the new bit 0
        data[0] <= #`DEL ^(data & `TAPS);
`endif

    end
end
endmodule     // LFSR
```

(handwritten margin notes: "mm)zone", "ver function Q and Taps 9xor 9nor")

13.2.2 Implementation 2

```
/************************************************************/
// MODULE:       linear feedback shift register (LFSR)
//
// FILE NAME:    lfsr2_rtl.v
// VERSION:      1.1
// DATE:         January 1, 2003
// AUTHOR:       Bob Zeidman, Zeidman Consulting
//
// CODE TYPE:    Register Transfer Level
//
// DESCRIPTION:  This module defines a linear feedback shift
// register (LFSR) using multiple XOR gates.
//
/************************************************************/
// DEFINES
`define DEL   1      // Clock-to-output delay. Zero
                     // time delays can be confusing
                     // and sometimes cause problems.
                     // These are good tap values for 2 to 32 bits
```

```
`define TAP2      2'b11
`define TAP3      3'b101
`define TAP4      4'b1001
`define TAP5      5'b10010
`define TAP6      6'b100001
`define TAP7      7'b1000001
`define TAP8      8'b10001110
`define TAP9      9'b100001000
`define TAP10     10'b1000000100
`define TAP11     11'b10000000010
`define TAP12     12'b100000101001
`define TAP13     13'b1000000001101
`define TAP14     14'b10000000010101
`define TAP15     15'b100000000000001
`define TAP16     16'b1000000000010110
`define TAP17     17'b10000000000000100
`define TAP18     18'b100000000001000000
`define TAP19     19'b1000000000000010011
`define TAP20     20'b10000000000000000100
`define TAP21     21'b100000000000000000010
`define TAP22     22'b1000000000000000000001
`define TAP23     23'b10000000000000000010000
`define TAP24     24'b100000000000000000001101
`define TAP25     25'b1000000000000000000000100
`define TAP26     26'b10000000000000000000100011
`define TAP27     27'b100000000000000000000010011
`define TAP28     28'b1000000000000000000000000100
`define TAP29     29'b10000000000000000000000000010
`define TAP30     30'b100000000000000000000000101001
`define TAP31     31'b1000000000000000000000000000100
`define TAP32     32'b10000000000000000000000001100010

`define BITS 8          // Number of bits in the LFSR
`define TAPS `TAP8       // This must be the taps for the
                        // number of bits specified above
`define INIT 1          // This can be any non-zero value
                        // for initialization of the LFSR

// TOP MODULE
module LFSR(
      clk,
      reset,
      data);

// PARAMETERS

// INPUTS
input             clk;      // Clock
```

```
input                 reset;      // Reset

// OUTPUTS
output [`BITS-1:0]   data;       // LFSR data

// INOUTS

// SIGNAL DECLARATIONS
wire                 clk;
wire                 reset;
reg [`BITS-1:0]      data;

// ASSIGN STATEMENTS

// MAIN CODE

// Look at the rising edge of clock or reset
always @(posedge clk or posedge reset) begin
    if (reset)
        data <= #`DEL `INIT;
    else begin
        // XOR each tap bit with the most significant bit
        // and shift left. Also XOR the most significant
        // bit with the incoming bit
        if (data[`BITS-1]) begin
            data <= #`DEL (data ^ `TAPS) << 1;
`ifdef ADD_ZERO  // Use this code if data == 0 is required
            data[0] <= #`DEL data[`BITS-1] ^ ~|(data ^ `TAPS);
`else
            data[0] <= #`DEL data[`BITS-1];
`endif

        end
        else begin
            data <= #`DEL data << 1;
`ifdef ADD_ZERO  // Use this code if data == 0 is required
            data[0] <= #`DEL data[`BITS-1] ^ ~|data;
`else
            data[0] <= #`DEL data[`BITS-1];
`endif

        end
    end
end
endmodule        // LFSR
```

13.3 SIMULATION CODE

The code for simulating the LFSR is shown below. Note that the code does not check that each value of the pseudo-random sequence is actually generated, but only that the initial value appears again after the correct number of cycles. This is all that is necessary to know that the LFSR has worked correctly. The reason for this is due to the fact that the LFSR next value relies only on its current value. Because there are only 2^n-1 possible values that the LFSR can possibly have, and we are checking 2^n-1 cycles, there are only two type of failures. The first failure is that one or more of the possible values is never included in the sequence. In this case, the cycle would have to repeat earlier than expected, a condition for which we check. The other failure is that the LFSR repeats numbers in the sequence. However, once a number is repeated, the entire sequence after that number would be repeated indefinitely. In this case, either the initialized value is in the repeated sequence, in which case the initial value shows up too early and is reported as an error. Otherwise the initial value is not in the sequence and at the end of the full number of cycles, when we do not have the initial value again, an error is reported. So all error conditions are covered with a minimum amount of code and checking.

Note that the same add_zero constant is used here as it is in the behavioral and RTL code, if you need to have 0 inserted into the sequence. In that case, define the constant, and the simulation will loop for one extra cycle to cover the inserted value.

```
/*******************************************************/
// MODULE:       LFSR simulation
//
// FILE NAME:    lfsr_sim.v
// VERSION:      1.1
// DATE:         January 1, 2003
// AUTHOR:       Bob Zeidman, Zeidman Consulting
//
// CODE TYPE:    Simulation
//
// DESCRIPTION:  This module provides stimuli for simulating
// a Linear Feedback Shift Register. This simulation cycles
// for the full number of cycles and checks that we get the
// initial value again exactly at the end of the sequence,
// not sooner or later.
//
/*******************************************************/


// DEFINES
`define DEL   1      // Clock-to-output delay. Zero
                     // time delays can be confusing
                     // and sometimes cause problems.

`define BITS 8       // Number of bits in the LFSR
```

```verilog
// TOP MODULE
module lfsr_sim();

// PARAMETERS

// INPUTS

// OUTPUTS

// INOUTS

// SIGNAL DECLARATIONS
reg             clock;
reg             reset;
wire [`BITS-1:0] data;

integer         cycle_count;    // Cycle count variable
integer         cycle_limit;    // The number of cycles to
                                // simulate
integer         init_value;     // Used to store the
                                // initial value of LFSR
// ASSIGN STATEMENTS

// MAIN CODE

// Instantiate the counter
LFSR lfsr(
     .clk(clock),
     .reset(reset),
     .data(data));

// Initialize inputs
initial begin
   clock = 1;
   reset = 0;
   cycle_count = 0;

`ifdef ADD_ZERO  // Use this code if data == 0 is required
   cycle_limit = (32'h00000001 << `BITS);
`else            // Use this code for a standard LFSR
   cycle_limit = (32'h00000001 << `BITS)-1;
`endif

   // Reset the LFSR and record the initial value
   #10 reset = 1;
   #10 reset = 0;
   init_value <= data;
end
```

```verilog
// Generate the clock
always #100 clock = ~clock;

// Simulate
always @(posedge clock) begin
    // Wait for outputs to settle
    #`DEL;
    #`DEL;

    // Increment the cycle count
    cycle_count = cycle_count + 1;

    // Check whether we have cycled back to the
    // original value
    if (data === init_value) begin
        if (cycle_count === cycle_limit) begin
            $display("\nSimulation complete - no errors\n");
            $finish;
        end
        else begin
            $display("\nERROR at time %0t:", $time);
            $display("LFSR cycled too quickly");
            $display("    initial value = %h", init_value);
            $display("    current value = %h", data);
            $display("    cycle count  = %d\n", cycle_count);

// Use $stop for debugging
            $stop;
        end
    end
    else if (cycle_count === cycle_limit) begin
        $display("\nERROR at time %0t:", $time);
        $display("LFSR should have cycled by now");
        $display("    initial value = %h", init_value);
        $display("    current value = %h", data);
        $display("    cycle count  = %d\n", cycle_count);

        // Use $stop for debugging
        $stop;
    end
end
endmodule    // lfsr_sim
```

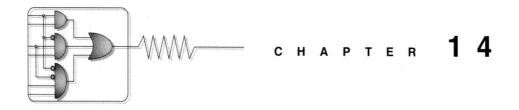

The Encrypter/Decrypter

*E*ncryption and decryption of data is an important application today, as growing amounts of sensitive data need to be kept secret. A simple form of encryption, known as private key encryption, can be implemented using a Linear Feedback Shift Register (LFSR) as described in the previous chapter. With this form of encryption, the key is a value used to initialize or "seed" the LFSR, which then produces a psuedo-random sequence of values. Each value is then combined with each word of data, through XOR gates, to produce the encrypted word. To decrypt the data, the LFSR is again seeded with the key, and again each value is combined with each word of encrypted data, through XOR gates, to produce the original data word. The hardware for encryption and decryption is identical, as shown in Figure 14-1 and Figure 14-2. The larger the data width, n, the more difficult it is to guess the encryption key, since there are 2^n-1 possible keys and only one will produce the correct sequence. In addition, there are multiple possibilities for a set of LFSR taps to produce a psuedo-random sequence.

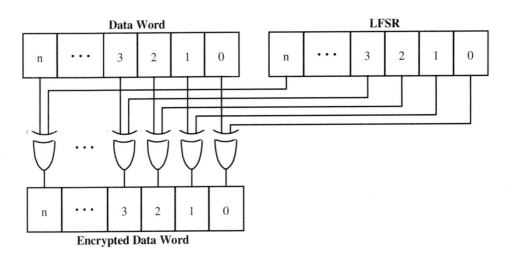

Figure 14-1 Encryption using an LFSR.

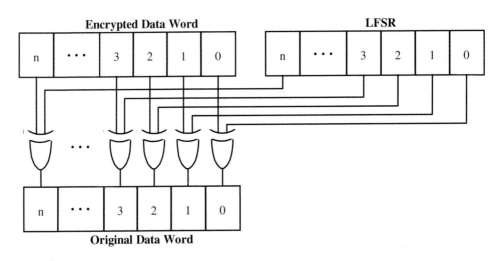

Figure 14-2 Decryption using an LFSR.

Public key encryption, which is more common in open network applications such as the Internet, involves the same basic technology. However, each user has a key that is known to everyone, hence the term "public" key. The sender and the recipient then exchange information about their keys in order to create private keys for encrypting and decrypting the data.

14.1 BEHAVIORAL CODE

```
/*******************************************************/
// MODULE:      encryption and decryption
//
// FILE NAME:   encr_beh.v
// VERSION:     1.1
// DATE:        January 1, 2003
// AUTHOR:      Bob Zeidman, Zeidman Consulting
//
// CODE TYPE:   Behavioral Level
//
// DESCRIPTION: This module defines a data
// encrypter/decrypter
//
/*******************************************************/

// DEFINES
`define DEL   1      // Clock-to-output delay. Zero
                     // time delays can be confusing
                     // and sometimes cause problems.

                     // These are good tap values for 2 to 32 bits
`define TAP2     2'b11
`define TAP3     3'b101
`define TAP4     4'b1001
`define TAP5     5'b10010
`define TAP6     6'b100001
`define TAP7     7'b1000001
`define TAP8     8'b10001110
`define TAP9     9'b100001000
`define TAP10    10'b1000000100
`define TAP11    11'b10000000010
`define TAP12    12'b100000101001
`define TAP13    13'b1000000001101
`define TAP14    14'b10000000010101
`define TAP15    15'b100000000000001
`define TAP16    16'b1000000000010110
`define TAP17    17'b10000000000000100
`define TAP18    18'b100000000001000000
`define TAP19    19'b1000000000000010011
`define TAP20    20'b10000000000000000100
`define TAP21    21'b100000000000000000010
`define TAP22    22'b1000000000000000000001
`define TAP23    23'b10000000000000000010000
`define TAP24    24'b100000000000000000001101
`define TAP25    25'b1000000000000000000000100
```

```verilog
`define TAP26    26'b10000000000000000000100011
`define TAP27    27'b100000000000000000000010011
`define TAP28    28'b1000000000000000000000000100
`define TAP29    29'b10000000000000000000000000010
`define TAP30    30'b100000000000000000000000101001
`define TAP31    31'b1000000000000000000000000000100
`define TAP32    32'b10000000000000000000000001100010
`define WORD_SZ 8    // Number of bits per word
`define TAPS `TAP8   // This must be the taps for the
                     // number of bits specified above

// TOP MODULE
module Encrypt(
      clk,
      load,
      key,
      data_in,
      data_out);

// PARAMETERS

// INPUTS
input                    clk;         // Clock
input                    load;        // Load the key
input [`WORD_SZ-1:0]     key;           // Encryption key
input [`WORD_SZ-1:0]     data_in;       // Input data

// OUTPUTS
output [`WORD_SZ-1:0]    data_out;    // Encrypted data

// INOUTS

// SIGNAL DECLARATIONS
wire                     clk;
wire                     load;
wire [`WORD_SZ-1:0]      key;
wire [`WORD_SZ-1:0]      data_in;
wire [`WORD_SZ-1:0]      data_out;

reg  [`WORD_SZ-1:0]      lfsr;        // LFSR data
reg  [`WORD_SZ-1:0]      temp;        // Temporary storage

// ASSIGN STATEMENTS
                                      // Encrypt the data
assign #`DEL data_out = lfsr ^ data_in;

// MAIN CODE
// Look for the load signal
always @(load) begin
```

```
    // if load gets deasserted, unforce the output, but
    // have it keep its current value
    if (~load) begin
       temp = key;
       deassign lfsr;
    end
    // Wait for the rising edge of the clock
    @(posedge clk);
    // If load is asserted at the clock, force the output
    // to the current input value (delayed by `DEL)
    if (load) begin
       temp = key;
       #`DEL assign lfsr = temp;
    end
end
// Look at the rising edge of clock for state transitions
always @(posedge clk) begin
    // Create the next random word for encryption

    // Shift all of the bits left
    lfsr[`WORD_SZ-1:1] <= #`DEL lfsr[`WORD_SZ-2:0];

    // Create the new bit 0
    lfsr[0] <= #`DEL ^(lfsr & `TAPS);
end
endmodule     // Encrypt
```

14.2 RTL CODE

```
/**********************************************************/
// MODULE:       encryption and decryption
//
// FILE NAME:    encr_rtl.v
// VERSION:      1.1
// DATE:         January 1, 2003
// AUTHOR:       Bob Zeidman, Zeidman Consulting
//
// CODE TYPE:    Register Transfer Level
//
// DESCRIPTION:  This module defines a data
// encrypter/decrypter
//
/**********************************************************/

// DEFINES
```

```
`define DEL   1       // Clock-to-output delay. Zero
                      // time delays can be confusing
                      // and sometimes cause problems.

                      // These are good tap values for 2 to 32 bits
`define TAP2      2'b11
`define TAP3      3'b101
`define TAP4      4'b1001
`define TAP5      5'b10010
`define TAP6      6'b100001
`define TAP7      7'b1000001
`define TAP8      8'b10001110
`define TAP9      9'b100001000
`define TAP10     10'b1000000100
`define TAP11     11'b10000000010
`define TAP12     12'b100000101001
`define TAP13     13'b1000000001101
`define TAP14     14'b10000000010101
`define TAP15     15'b100000000000001
`define TAP16     16'b1000000000010110
`define TAP17     17'b10000000000000100
`define TAP18     18'b100000000001000000
`define TAP19     19'b1000000000000010011
`define TAP20     20'b10000000000000000100
`define TAP21     21'b100000000000000000010
`define TAP22     22'b1000000000000000000001
`define TAP23     23'b10000000000000000010000
`define TAP24     24'b100000000000000000001101
`define TAP25     25'b1000000000000000000000100
`define TAP26     26'b10000000000000000000100011
`define TAP27     27'b100000000000000000000010011
`define TAP28     28'b1000000000000000000000000100
`define TAP29     29'b10000000000000000000000000010
`define TAP30     30'b100000000000000000000000101001
`define TAP31     31'b1000000000000000000000000000100
`define TAP32     32'b10000000000000000000000001100010

`define WORD_SZ 8   // Number of bits per word
`define TAPS `TAP8  // This must be the taps for the
                    // number of bits specified above

// TOP MODULE
module Encrypt(
      clk,
      load,
      key,
      data_in,
      data_out);
```

```
// PARAMETERS

// INPUTS
input                   clk;        // Clock
input                   load;       // Load the key
input [`WORD_SZ-1:0]    key;            // Encryption key
input [`WORD_SZ-1:0]    data_in;        // Input data

// OUTPUTS
output [`WORD_SZ-1:0]   data_out; // Encrypted data

// INOUTS

// SIGNAL DECLARATIONS
wire                    clk;
wire                    load;
wire [`WORD_SZ-1:0]     key;
wire [`WORD_SZ-1:0]     data_in;
wire [`WORD_SZ-1:0]     data_out;

reg  [`WORD_SZ-1:0]     lfsr;       // LFSR data

// ASSIGN STATEMENTS

                                    // Encrypt the data
assign #`DEL data_out = lfsr ^ data_in;
// MAIN CODE

// Look at the rising edge of clock for state transitions
always @(posedge clk) begin
    if (load) begin
        // Load the key into the LFSR
        lfsr <= #`DEL key;
    end
    else begin
        // Create the next random word for encryption

        // Shift all of the bits left
        lfsr[`WORD_SZ-1:1] <= #`DEL lfsr[`WORD_SZ-2:0];

        // Create the new bit 0
        lfsr[0] <= #`DEL ^(lfsr & `TAPS);
    end
end
endmodule     // Encrypt
```

14.3 SIMULATION CODE

This simulation code generates a random sequence of data words and a random encryption key. It uses the key to encrypt the sequence and store the encrypted data. Then it uses the key to decrypt the data and confirms that the decrypted data matches the original data.

```verilog
/*******************************************************/
// MODULE:         encrypter/decrypter simulation
//
// FILE NAME:      encr_sim.v
// VERSION:        1.2
// DATE:           January 1, 2003
// AUTHOR:         Bob Zeidman, Zeidman Consulting
//
// CODE TYPE:      Simulation
//
// DESCRIPTION:  This module provides stimuli for simulating
// a data encrypter/decrypter. It generates a random sequence
// of data words and a random encryption key. It uses the key
// to encrypt the sequence and store the encrypted data. Then
// it uses the key to decrypt the data and confirms that the
// decrypted data matches the original data.
//
/*******************************************************/

// DEFINES
`define DEL   1          // Clock-to-output delay. Zero
                         // time delays can be confusing
                         // and sometimes cause problems.
`define WORD_SZ 8        // Number of bits in a data word
`define SEQ_LEN 100      // Number of words to encrypt

// TOP MODULE
module encr_sim();

// PARAMETERS

// INPUTS

// OUTPUTS

// INOUTS

// SIGNAL DECLARATIONS
reg                clock;
reg                load;
reg  [`WORD_SZ-1:0] data_in;
reg  [`WORD_SZ-1:0] key;
```

```verilog
wire ['WORD_SZ-1:0] data_out;

reg  ['WORD_SZ-1:0] data_exp;        // Expected data output
                                     // Data memory
reg  ['WORD_SZ-1:0] data_mem[0:'SEQ_LEN-1];
integer             index;           // Index into data memory

reg                 encrypt_flag;    // Are we encrypting
                                     // or decrypting?

// ASSIGN STATEMENTS

// MAIN CODE

// Instantiate the encrypter
Encrypt encrypt(
        .clk(clock),
        .load(load),
        .key(key),
        .data_in(data_in),
        .data_out(data_out));

// Initialize inputs
initial begin
    clock = 0;
    load = 1;             // Load the key
    index = 0;
    encrypt_flag = 1;     // Encrypting

    key = $random(0);     // Initialize random number generator
    key = {$random} % ('WORD_SZ'h0 - 'WORD_SZ'h1);
end

// Generate the clock
always #100 clock = ~clock;

// Simulate
always @(negedge clock) begin
    if (index === 1) begin
        // Stop loading the key
        load = 0;
    end
    if (encrypt_flag) begin
        // Input random data for encryption
        data_in = {$random} % ('WORD_SZ'h0 - 'WORD_SZ'h1);
        // Wait for the outputs to settle
        #'DEL;
        #'DEL;
        #'DEL;      // Save the encrypted output data for later
```

```verilog
        data_mem[index] = data_out;

        // Is it the end of the sequence?
        if (index === `SEQ_LEN-1) begin
            // Start over and begin decrypting
            index = 0;
            encrypt_flag = 0;
            key = $random(0);    // Initialize random number generator
            load = 1;            // Reload the key into the circuit
            key = {$random} % (`WORD_SZ'h0 - `WORD_SZ'h1);
        end
        else begin
            // Increment the index
            index = index + 1;
        end
    end
    else begin
        // Decrypt the encrypted data
        data_in = data_mem[index];

        // Wait for the outputs to settle
        #`DEL;
        #`DEL;
        #`DEL;

        // Get the expected data
        data_exp = {$random} % (`WORD_SZ'h0 - `WORD_SZ'h1);

        // Check the output
        if (data_exp !== data_out) begin
            $display("\nERROR at time %0t:", $time);
            $display("Encryption/decryption not working");
            $display("    expected data = %h", data_exp);
            $display("    actual data   = %h\n", data_out);

            // Use $stop for debugging
            $stop;
        end       // Is is the end of the sequence?
        if (index === `SEQ_LEN-1) begin
            $display("\nSimulation complete - no errors\n");
            $finish;
        end
        else begin
            // Increment the index
            index = index + 1;
        end
    end
end
endmodule        // encr_sim
```

The Phase Locked Loop (PLL)

*T*he Phase Locked Loop (PLL) is particularly important in communication hardware designs. It is used to create a clock that is of a known frequency, from another clock that is not quite regular, or from a data stream that has the clock embedded in it. The digital PLL described here takes an input that tells it the frequency of the clock that it is supposed to generate. If the frequency is fixed, this can be coded as a constant, further optimizing the code. The PLL looks at an incoming signal to attempt to match its own edges to that of the incoming signal. This is how it locks onto the phase of the incoming clock. If the edges do not line up (i.e., the phases do not match), the PLL must attempt to adjust its own output to synchronize the clock phases. It does this slowly to minimize the jitter in its output, which is a term for how quickly the clock edges jump around, creating unwanted high frequency noise.

15.1 BEHAVIORAL CODE

The following behavioral code implements a PLL in an asynchronous way. This makes simulation much faster because the code is not evaluated on every system clock edge. However, there are many potential problems in asynchronous code, such as race conditions. In particular, care must be taken to avoid situations where code within different *always* blocks must be executed in a certain order for the function to work correctly.

This particular PLL function generates an output clock of the required frequency. When an output clock rising edge occurs, it stores the time for the next rising clock edge in the variable high_time and schedules clock_high to be loaded with the same time value at that time. For example, if the next rising edge is scheduled to occur at time 500, high_time will immediately be set to 500, while clock_high will be scheduled to get the value of 500 at time 500. When the input clock rising edge is sampled, if the output clock is determined to be out of phase, the code stores a new time value in high_time and schedules another change for clock_high. Now clock_high is scheduled to change at two different times. At the scheduled times for clock_high to change, if the time value stored in high_time matches the new value of clock_high, the output clock will change because this means that the last determination of the correct time for the output clock rising edge is at this current time. Otherwise, if the two values do not match, this means that this clock rising edge was later overridden by the circuit in order to change the phase of the output clock.

If that explanation was not clear to you, this table with hard numbers may make it easier to understand:

Table 15-1   PLL Behavioral Code with a Clock Period of 100 Time Units

Time	Event	high_time	Clock_high	output clock
10	output clock rising edge occurs	x	x	no change
	set value of high_time to the current time + 100 (110)	then 110		
	schedule clock_high to change to 110 at time 110			
110	output clock rising edge occurs	110	110	rising edge
	set value of high_time to the current time + 100 (210)	then 210		
	schedule clock_high to change to 210 at time 210			
120	*input* clock rising edge occurs	210	110	no change
	change value of high_time to the current time + 100 (220)	then 220		
	schedule clock_high to change to 220 at time 220			

Time	Event	high_time	Clock_high	output clock
	at time 220			
210	output clock does not change because the two variables have different values	220	210	no change
220	output clock changes because the two variables have the same values	220	220	rising edge

```
/********************************************************/
// MODULE:       Phase Locked Loop (PLL)
//
// FILE NAME:    pll_beh.v
// VERSION:      1.1
// DATE:         January 1, 2003
// AUTHOR:       Bob Zeidman, Zeidman Consulting
//
// CODE TYPE:    Behavioral Level
//
// DESCRIPTION:  This module defines a phase locked loop
// (PLL) with a synchronous reset. This PLL looks only at
// the rising edge of the input clock for synchronization.
// It is possible for a PLL to look at the falling edge or
// both edges also. Note that this code evaluates most
// inputs on the input and output clock edges, rather than
// the system clock. This speeds up execution since the input
// and output clocks are much slower than the system clock.
//
/********************************************************/

// DEFINES
`define DEL   1        // Clock-to-output delay. Zero
                       // time delays can be confusing
                       // and sometimes cause problems.
`define CNT_SZ 4       // The number of bits in the counter.
                       // This determines the maximum
                       // frequency of the PLL.
`define DUTY 2         // This determines the duty cycle of
                       // the output clock.
                       // 2 = 50% low, 50% high
                       // 3 = 33% low, 67% high
                       // 4 = 25% low, 75% high
                       // etc.
`define CYCLE_TIME 200 // System clock cycle time
```

```
// TOP MODULE
module PLL(
      reset,
      limit,
      clk,
      clk_in,
      clk_out);

// PARAMETERS

// INPUTS
input                 reset;    // We don't use this input,
                                // but keep it for compatibility
                                // with the RTL code
input [`CNT_SZ-1:0] limit;      // The upper limit for the counter
                                // which determines the output
                                // clock frequency. This must be
                                // close to the input clock
                                // frequency.
input                 clk;      // Fast system clock
input                 clk_in;   // Clock input

// OUTPUTS
output                clk_out;  // Clock output

// INOUTS

// SIGNAL DECLARATIONS
wire                  reset;
wire [`CNT_SZ-1:0]  limit;
wire                  clk;
wire                  clk_in;
reg                   clk_out;

time                  edge_time;    // Keep track of clock edge time
time                  period;       // Target output clock period
time                  clk_high;     // Used to store the most
                                    // up-to-date time value for the
                                    // output clock to go high
time                  high_time;    // This value changes at the
time
                                    // that the output clock is to go
                                    // high
// ASSIGN STATEMENTS

// MAIN CODE
```

```
// Initialize regs
initial begin
    // Wait for the clock frequency to be defined
    wait (period !== 64'hxxxxxxxxxxxxxxxx);

    // Wait for the system clock, then start the output clock
    // in order to synchronize it to the system clock
    @(posedge clk);
    #`DEL;
    clk_high = $time;
    high_time = $time;
end

// Generate the output clock
always @(clk_high) begin
    // Only generate a clock edge if this is the correct
    // time to do so.
    if (clk_high == high_time) begin
        // Store the time of this edge
        edge_time = $time;

        // Generate the output clock rising edge
        clk_out <= 1'b1;
        // Generate the output clock falling edge
        clk_out <= #(period - period/`DUTY) 1'b0;

        // Set up the next clock period
        high_time = $time + period;
        clk_high <= #period (high_time);
    end
end

// Look at the rising edge of the input clock
always @(posedge clk_in) begin
    // Look for the rising edge of the input clock which
    // ideally happens exactly one period after the previous
    // rising edge of the output clock. If the time to the
    // previous rising edge is one period, we have
    // synchronized. Also, if the time to the previous rising
    // edge is 0 we have synchronized. This is because we
    // never know which always block will be executed first.
    if (($time-edge_time > 1) && ($time-edge_time != period)) begin
        if (period-($time-edge_time) > `CYCLE_TIME/2) begin
            // The output clock edge came too late, so subtract
            // a system clock cycle to the output clock period.
```

```
            high_time = high_time - `CYCLE_TIME;
            clk_high <= #(high_time-$time) high_time;
        end
        else if (($time-edge_time) > `CYCLE_TIME/2) begin
            // The output clock edge came too early, so add a
            // system clock cycle to the output clock period.
            high_time = high_time + `CYCLE_TIME;
            clk_high <= #(high_time-$time) high_time;
        end
    end
end

// Look for changes in the target clock period
always @(limit) begin
    @(posedge clk)
        period <= #`DEL (limit+1)*`CYCLE_TIME;
end
endmodule        // PLL
```

15.2 RTL CODE

```
/***********************************************************/
// MODULE:        Phase Locked Loop (PLL)
//
// FILE NAME:     pll_rtl.v
// VERSION:       1.1
// DATE:          January 1, 2003
// AUTHOR:        Bob Zeidman, Zeidman Consulting
//
// CODE TYPE:     Register Transfer Level
//
// DESCRIPTION:  This module defines a phase locked loop
// (PLL) with a synchronous reset. This PLL looks only at
// the rising edge of the input clock for synchronization.
// It is possible for a PLL to look at the falling edge or
// both edges also.
//
/***********************************************************/

// DEFINES
`define DEL    1        // Clock-to-output delay. Zero
                        // time delays can be confusing
                        // and sometimes cause problems.
`define CNT_SZ 4        // The number of bits in the counter.
                        // This determines the maximum
```

```
                              // frequency of the PLL.
`define DUTY 2                 // This determines the duty cycle of
                              // the output clock.
                              // 2 = 50% low, 50% high
                              // 3 = 33% low, 67% high
                              // 4 = 25% low, 75% high
                              // etc.

// TOP MODULE
module PLL(
      reset,
      limit,
      clk,
      clk_in,
      clk_out);

// PARAMETERS

// INPUTS
input                reset;     // Reset the PLL
input [`CNT_SZ-1:0]  limit;     // The upper limit for the
                              // counter which determines
                              // the output clock frequency.
                              // This must be close to the
                              // input clock frequency.
input                clk;       // Fast system clock
input                clk_in;    // Clock input

// OUTPUTS
output               clk_out;   // Clock output

// INOUTS

// SIGNAL DECLARATIONS
wire                 reset;
wire [`CNT_SZ-1:0]   limit;
wire                 clk;
wire                 clk_in;
wire                 clk_out;
reg  [`CNT_SZ-1:0]   counter;   // Counter used to
                              // lock onto the clock
reg                  reg_in;    // Registered clock input

// ASSIGN STATEMENTS
assign #`DEL clk_out = (counter > limit/`DUTY) ? 1'b1 : 1'b0;

// MAIN CODE
```

```verilog
// Look at the rising edge of system clock
always @(posedge clk) begin
    if (reset) begin
        counter <= #`DEL 0;
    end
    else begin
        // Store the clock input for finding rising edges
        reg_in <= #`DEL clk_in;

        // Look for the rising edge of the input clock which
        // ideally happens at the same time as
        // counter == limit. Add or remove counter cycles to
        // synchronize its phase with that of the input clock.
        if ((!reg_in & clk_in) && (counter != limit)) begin
            if (counter < limit/2) begin
                // The output clock edge came too late, so
                // subtract two from the counter instead of the
                // usual one.
                if (counter == 1)
                    counter <= #`DEL limit;
                else if (counter == 0)
                    counter <= #`DEL limit - 1;
                else
                    counter <= #`DEL counter - 2;
            end
            // Otherwise add an extra cycle by not decrementing
            // the counter this cycle
        end
        else begin
            if (counter == 0)
                counter <= #`DEL limit;
            else
                counter <= #`DEL counter - 1;
        end
    end
end
endmodule      // PLL
```

15.3 SIMULATION CODE

The simulation code uses a counter to generate a clock input to the PLL. The counter is a down counter. Changing the upper limit of the counter causes the output clock frequency to change. By looking for a low value of the output clock on the previous cycle and a high value on the current cycle, we determine that there has been a rising edge on the output clock. When this rising edge occurs during the same cycle as a rising edge on the input clock, three times in a row, then we consider the PLL to have locked to the incoming clock. A rising edge on the input clock occurs when the counter reaches zero.

We attempt to get the PLL to lock onto several different frequencies by changing the counter limit. If all of the frequencies can be locked onto by the PLL, the simulation ends successfully. If the PLL does not lock within a certain large number of cycles, the simulation stops and reports an error. Note that the input clock is purposely a clock with a lopsided duty cycle. The simulation waveforms can be examined to see that the output clock is approximately a square wave and does not depend on the duty cycle of the input clock.

```verilog
/******************************************************/
// MODULE:        Phase Locked Loop (PLL)
//
// FILE NAME:     pll_sim.v
// VERSION:       1.1
// DATE:          January 1, 2003
// AUTHOR:        Bob Zeidman, Zeidman Consulting
//
// CODE TYPE:     Simulation
//
// DESCRIPTION: This module provides stimuli for simulating
// a Phase Locked Loop. It uses a down counter to generate a
// clock input. When the output rising edge occurs during the
// same cycle as an input rising edge, three times in a row,
// we consider the PLL to have locked. We attempt to get the
// PLL to lock onto several different frequencies by changing
// the counter limit. If all of the frequencies can be
// locked, the simulation ends successfully. If the PLL does
// not lock within a certain large number of cycles, the
// simulation stops and reports an error.
//
/******************************************************/
// DEFINES
`define DEL   1              // Clock-to-output delay. Zero
                             // time delays can be confusing
                             // and sometimes cause problems.
`define CNT_SZ 4             // The number of bits in the counter.
                             // This determines the maximum
                             // frequency of the PLL.

`define CYCLE_TIME 200       // System clock cycle time

// TOP MODULE
module pll_sim();

// PARAMETERS

// INPUTS

// OUTPUTS
```

```
// INOUTS
// SIGNAL DECLARATIONS
reg                 reset;
reg                 clk;
wire                clk_in;
wire                clk_out;

integer             counter;            // Counter used to generate
                                        // the input clock
reg [`CNT_SZ-1:0]   limit;              // Limit for counter to
                                        // generate input clock
integer             cycle_count;        // Cycle counter
integer             case_count;         // Counts each test case
                                        // until we are done
integer             lock_count;         // Counts each time the
                                        // clocks are locked
reg                 reg_in;             // Registered clock input

// ASSIGN STATEMENTS
// Create a very lopsided clock input signal and see if the
// output is a square wave.
assign #`DEL clk_in = (counter == 0) ? 1'b1 : 1'b0;

// MAIN CODE
// Instantiate the PLL
PLL pll(
        .reset(reset),
        .limit(limit),
        .clk(clk),
        .clk_in(clk_in),
        .clk_out(clk_out));

// Initialize regs
initial begin
    clk = 0;
    cycle_count = 0;
    case_count = 0;
    lock_count = 0;
    counter = 0;
    reset = 1;

    limit = 15;

    // Hold the synchronous reset for more than one clock
    // period
    #120 reset = 0;

end
```

```verilog
// Generate the clock
always #(`CYCLE_TIME/2) clk = ~clk;

// Simulate
always @(posedge clk) begin
    if (lock_count === 3) begin
        // We've locked
        $display(
            "PLL has locked onto clock after %d cycles\n:",
            cycle_count);

        case (case_count)
            0: begin
                limit = 10;
            end
            1: begin
                limit = 5;
            end
            2: begin
                limit = 7;
            end
            3: begin
                limit = 6;
            end
            4: begin
                limit = 12;
            end
            5: begin
                limit = 10;
            end
            6: begin
                $display("\nSimulation complete - no errors\n");
                $finish;
            end
        endcase
        lock_count = 0;
        counter <= #`DEL 0;
        case_count = case_count + 1;
        cycle_count = 0;
    end

    // Check for no clock output
    if (cycle_count >= 16*16) begin
        // We lost the lock
        $display("\nERROR at time %0t:", $time);
        $display("    Clock is not locking\n");
```

```verilog
        // Use $stop for debugging
        $stop;
    end

    // Decrement the counter
    if (counter === 0)
        counter <= #`DEL limit;
    else
        counter <= #`DEL counter - 1;
    // Increment the cycle count
    cycle_count = cycle_count + 1;
end

always @(negedge clk) begin
    // Look for the rising edge
    if (!reg_in & clk_out) begin
        if (counter === 0) begin
            // We got an edge correct
            lock_count = lock_count + 1;
        end
        else if (lock_count !== 0) begin
            // We lost the lock
            lock_count = 0;
        end
    end

    // Store the previous value of the clock
    reg_in <= clk_out;
end
endmodule        // pll_sim
```

CHAPTER 16

The Unsigned Integer Multiplier

The unsigned integer multiplier is a function that takes two unsigned integers as inputs and produces a result that is the product of the two integers. Seems simple. However, there are lots of ways of implementing this function, and each one has its advantages. I will give some different implementations so that you can choose the one that best meets your requirements and constraints. Each of these can be further optimized for space or speed.

16.1 BEHAVIORAL CODE

Ironically, while the RTL code for a multiplier can be very complex, the behavioral code is about as easy as it gets.

```
/***********************************************************/
// MODULE:          unsigned integer multiplier
//
// FILE NAME:       umultiply_beh.v
// VERSION:         1.1
// DATE:            January 1, 2003
// AUTHOR:          Bob Zeidman, Zeidman Consulting
//
// CODE TYPE:       Behavioral Level
//
// DESCRIPTION: This module defines a multiplier of
// unsigned integers.
//
/***********************************************************/

// DEFINES
`define DEL  1       // Clock-to-output delay. Zero
                     // time delays can be confusing
                     // and sometimes cause problems.
`define OP_BITS 4    // Number of bits in each operand

// TOP MODULE
module UnsignedMultiply(
     clk,
     reset,
     a,
     b,
     multiply_en,
     product,
     valid);

// PARAMETERS

// INPUTS
input                    clk;          // Clock
input                    reset;        // Reset input
input [`OP_BITS-1:0]     a;            // Multiplicand
input [`OP_BITS-1:0]     b;            // Multiplier
input                    multiply_en;  // Multiply enable input

// OUTPUTS
output [2*`OP_BITS-1:0]  product;      // Product
output                   valid;        // Is output valid yet?

// INOUTS

// SIGNAL DECLARATIONS
wire                     clk;
```

```
wire                        reset;

wire [`OP_BITS-1:0]         a;
wire [`OP_BITS-1:0]         b;
wire                        multiply_en;
reg  [2*`OP_BITS-1:0]    product;
reg                         valid;

// ASSIGN STATEMENTS

// MAIN CODE

// Look at the reset input
always @(reset) begin
    if (reset)
        #`DEL assign valid = 1'b0;
    else
        deassign valid;
end

// Look at the rising edge of the clock
always @(posedge clk) begin
    if (multiply_en) begin
        product <= #`DEL a*b;
        valid <= #`DEL 1'b1;
    end
end
endmodule     // UnsignedMultiply
```

16.2 RTL CODE

16.2.1 Brute Force Combinatorial Logic Implementation

The following implementation simply performs the entire multiplication using combinatorial logic. This RTL code is almost identical to the behavioral code. The advantage of this code is that it is simple to write. Also, it performs the entire operations in a single cycle. However, due to the large amount of combinatorial logic involved, it will probably slow your cycle time down significantly. If your logic is *extremely* fast, though, this would be a very efficient implementation.

```
/*********************************************************/
// MODULE:          unsigned integer multiplier
//
// FILE NAME:       umultiply1_rtl.v
```

```
// VERSION:         1.1
// DATE:            January 1, 2003
// AUTHOR:          Bob Zeidman, Zeidman Consulting
//
// CODE TYPE:       Register Transfer Level
//
// DESCRIPTION:  This module defines a multiplier of
// unsigned integers. This code uses a brute force
// combinatorial method for generating the product.
//
/*****************************************************/

// DEFINES
`define DEL    1      // Clock-to-output delay. Zero
                      // time delays can be confusing
                      // and sometimes cause problems.
`define OP_BITS 4    // Number of bits in each operand

// TOP MODULE
module UnsignedMultiply(
       clk,
       reset,
       a,
       b,
       multiply_en,
       product,
       valid);

// PARAMETERS

// INPUTS
input                     clk;            // Clock
input                     reset;          // Reset input
input [`OP_BITS-1:0]      a;              // Multiplicand
input [`OP_BITS-1:0]      b;              // Multiplier
input                     multiply_en;    // Multiply enable input

// OUTPUTS
output [2*`OP_BITS-1:0]   product;        // Product
output                    valid;          // Is the output valid?

// INOUTS

// SIGNAL DECLARATIONS
wire                      clk;
wire                      reset;
wire [`OP_BITS-1:0]       a;
wire [`OP_BITS-1:0]       b;
```

```
wire                        multiply_en;

reg  [2*`OP_BITS-1:0]      product;
reg                         valid;

// ASSIGN STATEMENTS

// MAIN CODE

// Look at the rising edge of the clock
// and the rising edge of reset
always @(posedge reset or posedge clk) begin
    if (reset)
        valid <= #`DEL 1'b0;
    else if (multiply_en)
        valid <= #`DEL 1'b1;
end


// Look at the rising edge of the clock
always @(posedge clk) begin
    if (multiply_en)
        product <= #`DEL a*b;
end
endmodule      // UnsignedMultiply
```

16.2.2 Brute Force ROM Implementation

The following code implements the multiplication in a ROM. The two operands are concatenated to form the address to the ROM. Each location in the ROM contains the product for those two operands. The advantage to this implementation is that it can be very fast. If the ROM access time is a single cycle, the multiplier can produce results each cycle. However, many technologies do not allow the implementation of ROMs. Also, large ROMs are often slow. If your target technology includes ROMs and they are fast, this is a very good implementation.

The actual ROM code is included at the bottom of this code in a separate file. For synthesis, you will need to know how to implement ROMs in the technology that you will be using to create your hardware. For simulation purposes, I have included a Verilog ROM in the next subsection. The following subsection after that shows the C code that can be used to create ROMs of any size for use in multipliers.

```
/************************************************************/
// MODULE:           unsigned integer multiplier
//
// FILE NAME:        umultiply2_rtl.v
```

```
// VERSION:          1.1
// DATE:             January 1, 2003
// AUTHOR:           Bob Zeidman, Zeidman Consulting
//
// CODE TYPE:        Register Transfer Level
//
// DESCRIPTION:  This module defines a multiplier of
// unsigned integers. This code uses a large ROM to generate
// each product.
//
/**********************************************************/

// DEFINES
`define DEL   1     // Clock-to-output delay. Zero
                    // time delays can be confusing
                    // and sometimes cause problems.
`define OP_BITS 4   // Number of bits in each operand

// TOP MODULE
module UnsignedMultiply(
      clk,
      reset,
      a,
      b,
      multiply_en,
      product,
      valid);

// PARAMETERS

// INPUTS
input                   clk;            // Clock
input                   reset;          // Reset input
input [`OP_BITS-1:0]    a;              // Multiplicand
input [`OP_BITS-1:0]    b;              // Multiplier
input                   multiply_en;    // Multiply enable input

// OUTPUTS
output [2*`OP_BITS-1:0] product;        // Product
output                  valid;          // Is the output valid?

// INOUTS

// SIGNAL DECLARATIONS
wire                    clk;
wire                    reset;
wire [`OP_BITS-1:0]     a:
```

```
wire [`OP_BITS-1:0]          b;
wire                         multiply_en;
reg  [2*`OP_BITS-1:0]        product;
reg                          valid;

wire [2*`OP_BITS-1:0]        romout;         // Output of ROM

// ASSIGN STATEMENTS

// MAIN CODE

// Instantiate the ROM
Um_Rom Rom(
    .address({a,b}),
    .out(romout));

// Look at the rising edge of the clock
// and the rising edge of reset
always @(posedge reset or posedge clk) begin
    if (reset)
        valid <= #`DEL 1'b0;
    else if (multiply_en)
        valid <= #`DEL 1'b1;
end

// Look at the rising edge of the clock
always @(posedge clk) begin
    if (multiply_en)
        product <= #`DEL romout;
end
endmodule      // UnsignedMultiply

// SUBMODULES
`include "um_rom.v"
```

16.2.2.1 Unsigned Multiplier ROM

The following code is an implementation of a ROM that is used to multiply two unsigned integers, each of which are 4 bits wide. As you can see, even this small ROM requires a lot of initialization code. That is why the next subsection has C code for automatically generating multiplier ROMs.

```
/*****************************************************************/
// MODULE:        Unsigned Integer Multiply ROM
//                for 4-bit operators
```

```
//
// FILE NAME:      um_rom.v
// VERSION:        1.1
// DATE:           Mon Apr 07 07:34:46 2003
// AUTHOR:         MR_GEN.EXE
//
// CODE TYPE:      Register Transfer Level
//
// DESCRIPTION:    This module defines a ROM that is used
// for generating the products of an unsigned integer
// multiplication. The operands are used as the index into
// the ROM and the output is the resulting product.
//
/**********************************************************/
// DEFINES

// TOP MODULE
module Um_Rom(
      address,
      out);

// PARAMETERS

// INPUTS
input [7:0]   address;    // ROM address

// OUTPUTS
output [7:0]  out;        // ROM output

// INOUTS

// SIGNAL DECLARATIONS
wire [7:0]    address;
wire [7:0]    out;

reg  [7:0]    rom[255:0];

// ASSIGN STATEMENTS
assign out = rom[address];


// MAIN CODE

initial begin
    rom[0] = 0;
    rom[1] = 0;
    rom[2] = 0;
```

```
rom[3]  = 0;
rom[4]  = 0;
rom[5]  = 0;
rom[6]  = 0;
rom[7]  = 0;
rom[8]  = 0;
rom[9]  = 0;
rom[10] = 0;
rom[11] = 0;
rom[12] = 0;
rom[13] = 0;
rom[14] = 0;
rom[15] = 0;
rom[16] = 0;
rom[17] = 1;
rom[18] = 2;
rom[19] = 3;
rom[20] = 4;
rom[21] = 5;
rom[22] = 6;
rom[23] = 7;
rom[24] = 8;
rom[25] = 9;
rom[26] = 10;
rom[27] = 11;
rom[28] = 12;
rom[29] = 13;
rom[30] = 14;
rom[31] = 15;
rom[32] = 0;
rom[33] = 2;
rom[34] = 4;
rom[35] = 6;
rom[36] = 8;
rom[37] = 10;
rom[38] = 12;
rom[39] = 14;
rom[40] = 16;
rom[41] = 18;
rom[42] = 20;
rom[43] = 22;
rom[44] = 24;
rom[45] = 26;
rom[46] = 28;
rom[47] = 30;
rom[48] = 0;
```

```
rom[49] = 3;
rom[50] = 6;
rom[51] = 9;
rom[52] = 12;
rom[53] = 15;
rom[54] = 18;
rom[55] = 21;
rom[56] = 24;
rom[57] = 27;
rom[58] = 30;
rom[59] = 33;
rom[60] = 36;
rom[61] = 39;
rom[62] = 42;
rom[63] = 45;
rom[64] = 0;
rom[65] = 4;
rom[66] = 8;
rom[67] = 12;
rom[68] = 16;
rom[69] = 20;
rom[70] = 24;
rom[71] = 28;
rom[72] = 32;
rom[73] = 36;
rom[74] = 40;
rom[75] = 44;
rom[76] = 48;
rom[77] = 52;
rom[78] = 56;
rom[79] = 60;
rom[80] = 0;
rom[81] = 5;
rom[82] = 10;
rom[83] = 15;
rom[84] = 20;
rom[85] = 25;
rom[86] = 30;
rom[87] = 35;
rom[88] = 40;
rom[89] = 45;
rom[90] = 50;
rom[91] = 55;
rom[92] = 60;
rom[93] = 65;
rom[94] = 70;
```

```
rom[95]  = 75;
rom[96]  = 0;
rom[97]  = 6;
rom[98]  = 12;
rom[99]  = 18;
rom[100] = 24;
rom[101] = 30;
rom[102] = 36;
rom[103] = 42;
rom[104] = 48;
rom[105] = 54;
rom[106] = 60;
rom[107] = 66;
rom[108] = 72;
rom[109] = 78;
rom[110] = 84;
rom[111] = 90;
rom[112] = 0;
rom[113] = 7;
rom[114] = 14;
rom[115] = 21;
rom[116] = 28;
rom[117] = 35;
rom[118] = 42;
rom[119] = 49;
rom[120] = 56;
rom[121] = 63;
rom[122] = 70;
rom[123] = 77;
rom[124] = 84;
rom[125] = 91;
rom[126] = 98;
rom[127] = 105;
rom[128] = 0;
rom[129] = 8;
rom[130] = 16;
rom[131] = 24;
rom[132] = 32;
rom[133] = 40;
rom[134] = 48;
rom[135] = 56;
rom[136] = 64;
rom[137] = 72;
rom[138] = 80;
rom[139] = 88;
rom[140] = 96;
```

```
rom[141] = 104;
rom[142] = 112;
rom[143] = 120;
rom[144] = 0;
rom[145] = 9;
rom[146] = 18;
rom[147] = 27;
rom[148] = 36;
rom[149] = 45;
rom[150] = 54;
rom[151] = 63;
rom[152] = 72;
rom[153] = 81;
rom[154] = 90;
rom[155] = 99;
rom[156] = 108;
rom[157] = 117;
rom[158] = 126;
rom[159] = 135;
rom[160] = 0;
rom[161] = 10;
rom[162] = 20;
rom[163] = 30;
rom[164] = 40;
rom[165] = 50;
rom[166] = 60;
rom[167] = 70;
rom[168] = 80;
rom[169] = 90;
rom[170] = 100;
rom[171] = 110;
rom[172] = 120;
rom[173] = 130;
rom[174] = 140;
rom[175] = 150;
rom[176] = 0;
rom[177] = 11;
rom[178] = 22;
rom[179] = 33;
rom[180] = 44;
rom[181] = 55;
rom[182] = 66;
rom[183] = 77;
rom[184] = 88;
rom[185] = 99;
rom[186] = 110;
```

```
rom[187] = 121;
rom[188] = 132;
rom[189] = 143;
rom[190] = 154;
rom[191] = 165;
rom[192] = 0;
rom[193] = 12;
rom[194] = 24;
rom[195] = 36;
rom[196] = 48;
rom[197] = 60;
rom[198] = 72;
rom[199] = 84;
rom[200] = 96;
rom[201] = 108;
rom[202] = 120;
rom[203] = 132;
rom[204] = 144;
rom[205] = 156;
rom[206] = 168;
rom[207] = 180;
rom[208] = 0;
rom[209] = 13;
rom[210] = 26;
rom[211] = 39;
rom[212] = 52;
rom[213] = 65;
rom[214] = 78;
rom[215] = 91;
rom[216] = 104;
rom[217] = 117;
rom[218] = 130;
rom[219] = 143;
rom[220] = 156;
rom[221] = 169;
rom[222] = 182;
rom[223] = 195;
rom[224] = 0;
rom[225] = 14;
rom[226] = 28;
rom[227] = 42;
rom[228] = 56;
rom[229] = 70;
rom[230] = 84;
rom[231] = 98;
rom[232] = 112;
rom[233] = 126;
rom[234] = 140;
```

```
        rom[235] = 154;
        rom[236] = 168;
        rom[237] = 182;
        rom[238] = 196;
        rom[239] = 210;
        rom[240] = 0;
        rom[241] = 15;
        rom[242] = 30;
        rom[243] = 45;
        rom[244] = 60;
        rom[245] = 75;
        rom[246] = 90;
        rom[247] = 105;
        rom[248] = 120;
        rom[249] = 135;
        rom[250] = 150;
        rom[251] = 165;
        rom[252] = 180;
        rom[253] = 195;
        rom[254] = 210;
        rom[255] = 225;
end
endmodule       // Um_Rom
```

16.2.2.2 C Code for Generating Multiplier ROMs

The following code is C code for generating a Verilog ROM for signed or unsigned multiplication. The usage of the code is:

```
mr_gen width filename signed
```

where
width	is the number of bits for each operand
filename	is the name of the output Verilog file
signed	is 0 for a ROM to implement unsigned multiplication
signed	is 1 for a ROM to implement signed multiplication

```
/*************************************************************/
/*************************************************************/
/*****                    MR_GEN                       *****/
/***** file: mr_gen.c                                  *****/
/*****                                                 *****/
/***** Written by Bob Zeidman                          *****/
/***** Zeidman Consulting                              *****/
/*****                                                 *****/
/***** Description:                                    *****/
```

```
/***** This program generates a Verilog ROM model for *****/
/***** an signed or unsigned multiplier. The arguments  *****/
/***** for the program are the width of each operand     *****/
/***** in bits, the output file name, and whether to     *****/
/***** generate a ROM for signed or unsigned             *****/
/***** multiplication.                                   *****/
/*****                                                   *****/
/***** Version Date        Author          Comments     *****/
/***** 1.0     8/17/98 Bob Zeidman   Original program  *****/
/***** 1.1     4/1/03 Bob Zeidman    Changed placement *****/
/*****             of parameters in                     *****/
/*****             Verilog code output                  *****/
/*****                                                  *****/
/***********************************************************/
/***********************************************************/

/***** Library Include Files *****/
/* ANSI C libraries */
#include <stdlib.h>              /* standard library */
#include <stdio.h>               /* standard io library */
#include <string.h>              /* string library */
#include <time.h>                /* time library */

/***** Global variables *****/

/***** Constants *****/
#define VERSION "1.1"            /* Program version */
#define FILELEN 76               /* Maximum number of characters
                                    in a file name */

/***** SUBROUTINES *****/

/***** Describe the usage of the program for the user *****/
void describe()
{
    printf("\nMultiplier ROM Generator Version %s\n",
        VERSION);
    printf("This program creates a Verilog ROM ");
    printf("for use with a multipler.\n");
    printf("\nUsage:\n");
    printf("\n\tmr_gen width filename s\n\n");
    printf("where width is the bit width of each ");
    printf("operator.\n");
    printf("        filename is the name of the file to ");
    printf("generate.\n");
    printf("        s = 0 for unsigned multiplication\n");
    printf("        s = 1 for signed multiplication\n");
}
```

```c
/***** MAIN ROUTINE *****/
/***** argc       Argument count              *****/
/***** argv[] Argument variables         *****/
void main(int argc, char *argv[])
{
    int width;              /* width of operands in bits */
    long index;             /* index counter for ROM */
    long last_index;        /* last index of ROM */
    long product;           /* product of two integers */
    int sign;               /* signed or unsigned multiply? */
    int a;                  /* multiplicand */
    int asign;              /* sign of a operand
                               (0 = positive, 1 = negative) */
    int b;                  /* multiplier */
    int bsign;              /* sign of b operand
                               (0 = positive, 1 = negative) */
    char filename[FILELEN]; /* Verilog file name */
    FILE *VerilogFile;      /* Verilog output file */
    time_t filetime;        /* Output file creation time */
    /* Check for exactly three arguments */
    if (argc != 4)
    {
        /* if there's not three arguments, it's wrong */
        /* describe the correct usage of this program,
           then exit */
        describe();
        exit(1);
    }

    /* Convert the first argument to an integer width */
    width = atoi(argv[1]);
    if (width <= 0)
    {
        /* if the width is not a whole number, it's wrong */
        /* describe the correct usage of this program,
           then exit */
        describe();
        exit(1);
    }

    /* Get the filename */
    strncpy(filename, argv[2], FILELEN-1);

    /* Open the file for writing */
    VerilogFile = fopen(filename, "wt");
    if (VerilogFile == NULL)
    {
```

```
        printf("Error: Cannot open file %s for writing!\n",
            filename);
        exit(1);
    }

    /* Convert the third argument to an integer */
    sign = atoi(argv[3]);
    if ((sign != 0) && (sign != 1))
    {
        /* if the sign is not 0 or 1, it's wrong */
        /* describe the correct usage of this program,
            then exit */
        describe();
        exit(1);
    }
    /* Make the file creation date string */
    time (&filetime);

    /* Calculate the last index into the ROM */
    last_index = (1 << 2*(width)) - 1;

    fprintf(VerilogFile, "/****************************");
    fprintf(VerilogFile, "***************************/\n");
    if (sign)
        fprintf(VerilogFile, "// MODULE:          Signed ");
    else
        fprintf(VerilogFile, "// MODULE:          Unsigned ");
    fprintf(VerilogFile, "Integer Multiply ROM\n");
    fprintf(VerilogFile, "//                   for %d-bit ",
        width);
    fprintf(VerilogFile, "operators\n");
    fprintf(VerilogFile, "//\n");
    fprintf(VerilogFile, "// FILE NAME:       %s\n", filename);
    fprintf(VerilogFile, "// VERSION:         %s\n", VERSION);
    fprintf(VerilogFile, "// DATE:            %s",
        ctime(&filetime));
    fprintf(VerilogFile, "// AUTHOR:          MR_GEN.EXE\n");
    fprintf(VerilogFile, "//\n");
    fprintf(VerilogFile, "// CODE TYPE:");
    fprintf(VerilogFile, "       Register Transfer Level\n");
    fprintf(VerilogFile, "//\n");
    fprintf(VerilogFile, "// DESCRIPTION:     This module ");
    fprintf(VerilogFile, "defines a ROM that is used\n");
    if (sign)
    {
        fprintf(VerilogFile, "// for generating the products");
        fprintf(VerilogFile, " of a signed integer\n");
    }
```

```
else
{
    fprintf(VerilogFile, "// for generating the products");
    fprintf(VerilogFile, " of an unsigned integer\n");
}
fprintf(VerilogFile, "// multiplication. The operands ");
fprintf(VerilogFile, "are used as the index into\n");
fprintf(VerilogFile, "// the ROM and the output is the ");
fprintf(VerilogFile, "resulting product.\n");
fprintf(VerilogFile, "//\n");
fprintf(VerilogFile, "/******************************");
fprintf(VerilogFile, "***************************/\n");
fprintf(VerilogFile, "\n");
fprintf(VerilogFile, "// DEFINES\n");
fprintf(VerilogFile, "\n");
fprintf(VerilogFile, "// TOP MODULE\n");
if (sign)
    fprintf(VerilogFile, "module Sm_Rom(\n");
else
    fprintf(VerilogFile, "module Um_Rom(\n");
fprintf(VerilogFile, "        address,\n");
fprintf(VerilogFile, "        out);\n");
fprintf(VerilogFile, "\n");
fprintf(VerilogFile, "// PARAMETERS\n");
fprintf(VerilogFile, "\n");
fprintf(VerilogFile, "// INPUTS\n");
fprintf(VerilogFile, "input [%d:0]   address", 2*width-1);
fprintf(VerilogFile, ";   // ROM address\n");
fprintf(VerilogFile, "\n");
fprintf(VerilogFile, "// OUTPUTS\n");
fprintf(VerilogFile, "output [%d:0]   out; ", 2*width-1);
fprintf(VerilogFile, "       // ROM output\n");
fprintf(VerilogFile, "\n");
fprintf(VerilogFile, "// INOUTS\n");
fprintf(VerilogFile, "\n");
fprintf(VerilogFile, "// SIGNAL DECLARATIONS\n");
fprintf(VerilogFile, "wire [%d:0]   address;\n",
    2*width-1);
fprintf(VerilogFile, "wire [%d:0]   out;\n", 2*width-1);
fprintf(VerilogFile, "\n");
fprintf(VerilogFile, "reg  [%d:0]   rom[%d:0];\n",
    2*width-1, last_index);
fprintf(VerilogFile, "\n");
fprintf(VerilogFile, "// ASSIGN STATEMENTS\n");
fprintf(VerilogFile, "assign out = rom[address];\n");
fprintf(VerilogFile, "\n");
fprintf(VerilogFile, "// MAIN CODE\n");
fprintf(VerilogFile, "\n");
```

```c
    fprintf(VerilogFile, "initial begin\n");
    /* Generate the ROM initialization */
    for (index = 0; index <= last_index; index++)
    {
        /* initialize the operand signs to positive */
        asign = 0;
        bsign = 0;
        a = (int)(index & ((1 << width) - 1));
        b = (int)(index >> width);
        if (sign)
        {
            if (a >> (width-1))
            {
                /* a is negative, so negate it
                   and remember the sign */
                asign = 1;
                a = (~a & ((1 << width) - 1)) + 1;
            }
            if (b >> (width-1))
            {
                /* b is negative, so negate it
                   and remember the sign */
                bsign = 1;
                b = (~b & ((1 << width) - 1)) + 1;
            }
        }
        product = a*b;
        if (asign^bsign)
        {
            /* if one operand is negative, but not both,
               the result needs to be negated */
            product = (~product & ((1 << 2*width) - 1)) + 1;
        }
        fprintf(VerilogFile, "\trom[%ld] = %ld;\n",
            index, product);
    }
    fprintf(VerilogFile, "end\n");
    if (sign)
        fprintf(VerilogFile, "endmodule        // Sm_Rom\n");
    else
        fprintf(VerilogFile, "endmodule        // Um_Rom\n");
}
```

16.2.3 Partial Sums Algorithm Implementation

The following Verilog code implements unsigned integer multiplication using what is called the partial sums algorithm. What is the partial sums algorithm? It is simply what we all do when we multiply using pencil and paper. For example, take a look at the multiplication problem shown in Figure 16-1. It is a simple decimal multiplication on paper. Figure 16-2 shows the same problem represented in binary. The process is identical, except in binary it is slightly easier. For each bit of the multiplier, if it is 1 we write down the multiplicand under it. If it is 0 we do nothing.

			5	6	multiplicand
x	1	3	5		multiplier

		2	8	0	partial sum
	1	6	8		partial sum
	5	6			partial sum

	7	5	6	0	product

Figure 16-1 Decimal multiplication of unsigned integers using partial sums.

Converting this algorithm to Verilog is fairly easy. We loop for each bit of the multiplier, starting with the most significant one. If the multiplier bit is a 1, we add the multiplicand to the product. If the bit is a 0, we do nothing to the product. In either case, we finish the cycle by shifting the product one bit to the left.

Note that I chose to start with the most significant bit, rather than the least significant bit as we would with paper and pencil. This is because starting with the most significant bit allows us to shift the product register. If we started with the least significant bit, we would need to store the multiplicand in a register and shift that register, requiring the use of this additional register.

						1	1	1	0	0	0	multiplicand
		x	1	0	0	0	0	1	1	1			multiplier

				1	1	1	0	0	0				partial sum
			1	1	1	0	0	0					partial sum
		1	1	1	0	0	0						partial sum
1	1	1	0	0	0								partial sum
	--											
1	1	1	0	1	1	0	0	0	1	0	0	0	product

Figure 16-2 Binary multiplication of unsigned integers using partial sums.

```
/*************************************************************/
// MODULE:          unsigned integer multiplier
//
// FILE NAME:       umultiply3_rtl.v
// VERSION:         1.1
// DATE:            January 1, 2003
// AUTHOR:          Bob Zeidman, Zeidman Consulting
//
// CODE TYPE:       Register Transfer Level
//
// DESCRIPTION:     This module defines a multiplier of
// unsigned integers. This code uses a shift register and
// partial sums.
//
/*************************************************************/

// DEFINES
`define DEL   1       // Clock-to-output delay. Zero
                      // time delays can be confusing
                      // and sometimes cause problems.
`define OP_BITS 4     // Number of bits in each operand
`define OP_SIZE 2     // Number of bits required to
                      // represent the OP_BITS constant

// TOP MODULE
module UnsignedMultiply(
      clk,
      reset,
      a,
      b,
      multiply_en,
      product,
      valid);


// PARAMETERS

// INPUTS
input                      clk;          // Clock
input                      reset;        // Reset input
input [`OP_BITS-1:0]   a;                // Multiplicand
input [`OP_BITS-1:0]   b;                // Multiplier
input                      multiply_en;  // Multiply enable input

// OUTPUTS
output [2*`OP_BITS-1:0]   product;       // Product
output                     valid;        // Is the output valid?

// INOUTS
```

```
// SIGNAL DECLARATIONS
wire                        clk;
wire                        reset;
wire  [`OP_BITS-1:0]        a;
wire  [`OP_BITS-1:0]        b;
wire                        multiply_en;
reg   [2*`OP_BITS-1:0]      product;
wire                        valid;

reg   [`OP_SIZE-1:0]          count;              // Count the number
                                                 // of bit shifts


// ASSIGN STATEMENTS
assign valid = (count == 0) ? 1'b1 : 1'b0;
// MAIN CODE

// Look at the rising edge of the clock
// and the rising edge of reset
always @(posedge reset or posedge clk) begin
    if (reset) begin
        count <= #`DEL `OP_SIZE'b0;
    end
    else begin
        if (multiply_en && valid) begin
            // Put the count at all ones - the maximum
            count <= #`DEL ~`OP_SIZE'b0;
        end
        else if (count)
            count <= #`DEL count - 1;
    end
end

// Look at the rising edge of the clock
always @(posedge clk) begin
    if (multiply_en & valid) begin
        if (b[`OP_BITS-1]) begin
            // If the MSB of the multiplier is 1,
            // load the multiplicand into the product
            product <= #`DEL a;
        end
        else begin
            // If the MSB of the multiplier is 0,
            // load 0 into the product
            product <= #`DEL 0;
        end
    end
    else if (count) begin
```

```
        if (b[count-1]) begin
            // If this bit of the multiplier is 1,
            // shift the product left and add the
            // multiplicand
            product <= #`DEL (product << 1) + a;
        end
        else begin
            // If this bit of the multiplier is 0,
            // just shift the product left
            product <= #`DEL product << 1;
        end
    end
end
endmodule    // UnsignedMultiply
```

16.3 SIMULATION CODE

The simulation code generates every possible combination of operands and examines each product for correctness.

```
/********************************************************/
// MODULE:          unsigned integer multiplier simulation
//
// FILE NAME:       umultiply_sim.v
// VERSION:         1.1
// DATE:            January 1, 2003
// AUTHOR:          Bob Zeidman, Zeidman Consulting
//
// CODE TYPE:       Simulation
//
// DESCRIPTION:   This module provides stimuli for simulating
// an unsigned integer multiplier. It generates every
// possible combination of operands and examines each product
// for correctness.
//
/********************************************************/

// DEFINES
`define OP_BITS 4 // Number of bits in each operand

// TOP MODULE
module umultiply_sim();

// PARAMETERS

// INPUTS
```

```
// OUTPUTS

// INOUTS

// SIGNAL DECLARATIONS
reg                         clock;
reg                         reset;
reg  [`OP_BITS-1:0]     a_in;
reg  [`OP_BITS-1:0]     b_in;
reg                         multiply_en;
wire [2*`OP_BITS-1:0] product_out;
wire                        valid;

reg  [2*`OP_BITS:0]     cycle_count;      // Counts valid clock cycles
integer                 val_count;      // Counts cycles
                                        // between valid data
reg  [2*`OP_BITS-1:0] expect;          // Expected output

// ASSIGN STATEMENTS

// MAIN CODE

// Instantiate the multiplier
UnsignedMultiply umult(
        .clk(clock),
        .reset(reset),
        .a(a_in),
        .b(b_in),
        .multiply_en(multiply_en),
        .product(product_out),
        .valid(valid));

// Initialize inputs
initial begin
   clock = 1;
   cycle_count = 0;

   reset = 1;           // Toggle reset to initialize
   #10 reset = 0;       // the valid output

   multiply_en = 1;     // Begin the operation

   val_count = 0;       // How many cycles to output valid data?
end
```

```verilog
// Generate the clock
always #100 clock = ~clock;

// Simulate
always @(negedge clock) begin
    if (valid === 1'b1) begin
        // Check the result for correctness
        if (product_out !== expect) begin
            $display("\nERROR at time %0t:", $time);
            $display("Adder is not working");
            if (multiply_en)
                $display("Multiplier is enabled");
            else
                $display("Multiplier is not enabled");
            $display("    a_in = %h", a_in);
            $display("    b_in = %h", b_in);
            $display("    expected result = %h",
                        expect);
            $display("    actual output = %h\n",
                        product_out);

            // Use $stop for debugging
            $stop;
        end

        // Create inputs between 0 and all 1s
        a_in = cycle_count[2*`OP_BITS-1:`OP_BITS];
        b_in = cycle_count[`OP_BITS-1:0];

        // How many cycles to output valid data?
        val_count = 0;

        // Count the valid cycles
        cycle_count = cycle_count + 1;
        if (cycle_count[2*`OP_BITS]) begin
            // We've cycled once
            case (cycle_count[1:0])
                0: begin
                    expect = a_in * b_in;
                end
                1: begin
                    multiply_en = 0; // Stop the operation

                    // We changed the inputs, but don't change the
                    // previous expected value since we have
                    // stopped the operation
                end
                2: begin
```

```
                $display("\nSimulation complete - no errors\n");
                $finish;
            end
        endcase
    end
    else begin
        expect = a_in * b_in;
    end
end
else begin
    // Keep track of how many cycles to output valid data
    val_count = val_count + 1;

    if (val_count > 2*`OP_BITS+3) begin
        $display("\nERROR at time %0t:", $time);
        $display("Too many cycles for valid data\n");

        // Use $stop for debugging
        $stop;
    end
end
end
endmodule        // umultiply_sim
```

The Signed Integer Multiplier

The signed integer multiplier is a function that takes two signed integers as inputs and produces a result thatis the product of the two integers. This function is slightly more complicated than the unsigned integer multiplier because it must take the sign into account. I have assumed that we are using two's complement representation of signed integers, because it is the most common, and most efficient, representation. As with the unsigned integer multiplier, there are lots of ways of implementing this function, and each one has its advantages. I will give some different implementations so that you can choose the one that best meets your requirements and constraints. Each of these can be further optimized for space or speed.

17.1 BEHAVIORAL CODE

As with the unsigned integer multiplier, the behavioral code for a signed integer multiplier is fairly easy. This model keeps track of the absolute values and the signs of the operands. It multiplies the absolute values and changes the sign of the product if necessary, based on the signs of the operands.

```
/************************************************************/
// MODULE:        signed integer multiplier
//
// FILE NAME:     smultiply_beh.v
// VERSION:       1.1
// DATE:          January 1, 2003
// AUTHOR:        Bob Zeidman, Zeidman Consulting
//
// CODE TYPE:     Behavioral Level
//
// DESCRIPTION:   This module defines a multiplier of signed
// integers.
//
/************************************************************/

// DEFINES
`define DEL   1        // Clock-to-output delay. Zero
                       // time delays can be confusing
                       // and sometimes cause problems.
`define OP_BITS 4      // Number of bits in each operand

// TOP MODULE
module SignedMultiply(
      clk,
      reset,
      a,
      b,
      multiply_en,
      product,
      valid);

// PARAMETERS

// INPUTS
input                      clk;          // Clock
input                      reset;        // Reset input
input [`OP_BITS-1:0]       a;            // Multiplicand
input [`OP_BITS-1:0]       b;            // Multiplier
input                      multiply_en;  // Multiply enable input

// OUTPUTS
output [2*`OP_BITS-1:0]    product;      // Product
output                     valid;        // Is output valid yet?
```

```
// INOUTS

// SIGNAL DECLARATIONS
wire                    clk;
wire                    reset;
wire [`OP_BITS-1:0]     a;
wire [`OP_BITS-1:0]     b;
wire                    multiply_en;
reg  [2*`OP_BITS-1:0]   product;
reg                     valid;
reg  [`OP_BITS-1:0]     a_abs;        // Absolute value of a
reg  [`OP_BITS-1:0]     b_abs;        // Absolute value of b

// ASSIGN STATEMENTS

// MAIN CODE

// Look at reset
always @(reset) begin
    if (reset)
        #`DEL assign valid = 0;
    else
        deassign valid;
end

// Look at the rising edge of the clock
always @(posedge clk) begin
    if (multiply_en) begin
        // Look at the sign bits, then multiply
        // their absolute values and negate the
        // result if necessary
        case ({a[`OP_BITS-1],b[`OP_BITS-1]})
            2'b00: begin
                product <= #`DEL a*b;
            end
            2'b01: begin
                b_abs = ~b + 1;
                product <= #`DEL ~(a*b_abs)+1;
            end
            2'b10: begin
                a_abs = ~a + 1;
                product <= #`DEL ~(a_abs*b)+1;
            end
            2'b11: begin
                a_abs = ~a + 1;
                b_abs = ~b + 1;
                product <= #`DEL a_abs*b_abs;
            end
```

```
        endcase
        valid <= #`DEL 1'b1;
    end
end
endmodule      // SignedMultiply
```

17.2 RTL CODE

17.2.1 Brute Force Combinatorial Logic Implementation

The following implementation simply performs the entire multiplication using combinatorial logic. This RTL code is almost identical to the behavioral code. The advantage of this code is that it is simple to write. Also, it performs the entire operations in a single cycle. However, due to the large amount of combinatorial logic involved, it will probably slow your cycle time down significantly. If your logic is *extremely* fast, though, this would be a very efficient implementation.

```
/*********************************************************/
// MODULE:          signed integer multiplier
//
// FILE NAME:       smultiply1_rtl.v
// VERSION:         1.0
// DATE:            January 1, 1999
// AUTHOR:          Bob Zeidman, Zeidman Consulting
//
// CODE TYPE:       Register Transfer Level
//
// DESCRIPTION:  This module defines a multiplier of signed
// integers.  This code uses a brute force combinatorial
// method for generating the product.
//
/*********************************************************/

// DEFINES
`define DEL   1      // Clock-to-output delay. Zero
                     // time delays can be confusing
                     // and sometimes cause problems.
`define OP_BITS 4    // Number of bits in each operand

// TOP MODULE
module SignedMultiply(
        clk,
        reset,
        a,
        b,
```

```
                    multiply_en,
                    product,
                    valid);

// PARAMETERS

// INPUTS
input                       clk;            // Clock
input                       reset;          // Reset input
input ['OP_BITS-1:0]        a;              // Multiplicand
input ['OP_BITS-1:0]        b;              // Multiplier
input                       multiply_en;    // Multiply enable input

// OUTPUTS
output [2*'OP_BITS-1:0]     product;        // Product
output                      valid;          // Is output valid yet?
// INOUTS

// SIGNAL DECLARATIONS
wire                        clk;
wire                        reset;
wire ['OP_BITS-1:0]         a;
wire ['OP_BITS-1:0]         b;
wire                        multiply_en;
reg  [2*'OP_BITS-1:0]       product;
reg                         valid;

reg  ['OP_BITS-1:0]         a_abs;          // Absolute value of a
reg  ['OP_BITS-1:0]         b_abs;          // Absolute value of b

// ASSIGN STATEMENTS

// MAIN CODE

// Look at the rising edge of the clock
// and the rising edge of reset
always @(posedge reset or posedge clk) begin
    if (reset)
        valid <= #'DEL 1'b0;
    else if (multiply_en)
        valid <= #'DEL 1'b1;
end

// Look at the rising edge of the clock
always @(posedge clk) begin
    if (multiply_en) begin
        // Look at the sign bits, then multiply
        // their absolute values and negate the
```

```
          // result if necessary
      case ({a[`OP_BITS-1],b[`OP_BITS-1]})
         2'b00: begin
            product <= #`DEL a*b;
         end
         2'b01: begin
            b_abs = ~b + 1;
            product <= #`DEL ~(a*b_abs)+1;
         end
         2'b10: begin
            a_abs = ~a + 1;
            product <= #`DEL ~(a_abs*b)+1;
         end
         2'b11: begin
            a_abs = ~a + 1;
            b_abs = ~b + 1;
            product <= #`DEL a_abs*b_abs;
         end
      endcase
   end
end
endmodule    // SignedMultiply
```

17.2.2 Brute Force ROM Implementation

The following code implements the multiplication in a ROM. The two operands are concatenated to form the address to the ROM. Each location in the ROM contains the product for those two operands. The advantage to this implementation is that it can be very fast. If the ROM access time is a single cycle, the multiplier can produce results each cycle. However, many technologies do not allow the implementation of ROMs. Also, large ROMs are often slow. If your target technology includes ROMs and they are fast, this is a very good implementation.

The actual ROM code is included at the bottom of this code in a separate file. For synthesis, you will need to know how to implement ROMs in the technology that you will be using to create your hardware. For simulation purposes, I have included a Verilog ROM in the next subsection. The previous chapter showed the C code that can be used to create ROMs of any size for use in multipliers.

```
/**************************************************************/
// MODULE:        signed integer multiplier
//
// FILE NAME:     smultiply1_rtl.v
// VERSION:       1.1
// DATE:          January 1, 2003
```

```
// AUTHOR:        Bob Zeidman, Zeidman Consulting
//
// CODE TYPE:     Register Transfer Level
//
// DESCRIPTION:  This module defines a multiplier of signed
// integers. This code uses a large ROM to generate each
// product.
//
/*****************************************************************/

// DEFINES
`define DEL   1       // Clock-to-output delay. Zero
                      // time delays can be confusing
                      // and sometimes cause problems.
`define OP_BITS 4     // Number of bits in each operand

// TOP MODULE
module SignedMultiply(
      clk,
      reset,
      a,
      b,
      multiply_en,
      product,
      valid);

// PARAMETERS

// INPUTS
input                      clk;                // Clock
input                      reset;              // Reset input
input [`OP_BITS-1:0]        a;                  // Multiplicand
input [`OP_BITS-1:0]        b;                  // Multiplier
input                      multiply_en;        // Multiply enable input

// OUTPUTS
output [2*`OP_BITS-1:0]     product;            // Product
output                     valid;              // Is output valid yet?
// INOUTS

// SIGNAL DECLARATIONS
wire                       clk;
wire                       reset;
wire [`OP_BITS-1:0]        a;
wire [`OP_BITS-1:0]        b;
wire                       multiply_en;
reg  [2*`OP_BITS-1:0]      product;
reg                        valid;
```

```
wire [2*`OP_BITS-1:0]        romout;

// Output of ROM

// ASSIGN STATEMENTS

// MAIN CODE

// Instantiate the ROM
Sm_Rom Rom(
    .address({a,b}),
    .out(romout));

// Look at the rising edge of the clock
// and the rising edge of reset
always @(posedge reset or posedge clk) begin
    if (reset)
        valid <= #`DEL 1'b0;
    else if (multiply_en)
        valid <= #`DEL 1'b1;
end

// Look at the rising edge of the clock
always @(posedge clk) begin
    if (multiply_en)
        product <= #`DEL romout;
end
endmodule      // SignedMultiply

// SUBMODULES
`include "sm_rom.v"
```

17.2.2.1 Signed Multiplier ROM

The following code is an implementation of a ROM that is used to multiply two signed
integers, each of which are 4 bits wide. As you can see, even this small ROM requires a lot of
initialization code. This can be done automatically using the C code in the previous chapter.

```
/***********************************************************/
// MODULE:        Signed Multiply ROM
//                for 4-bit operators
//
// FILE NAME:     sm_rom.v
```

```
// VERSION:        1.1
// DATE:           Mon Apr 07 07:35:04 2003
// AUTHOR:         MR_GEN.EXE
//
// CODE TYPE:      Register Transfer Level
//
// DESCRIPTION:    This module defines a ROM that is used
// for generating the products of a signed integer
// multiplication. The operands are used as the index into
// the ROM and the output is the resulting product.
//.
/*********************************************************/

// DEFINES

// TOP MODULE
module Sm_Rom(
      address,
      out);

// PARAMETERS

// INPUTS
input [7:0]    address;    // ROM address

// OUTPUTS
output [7:0]   out;        // ROM output

// INOUTS

// SIGNAL DECLARATIONS
wire [7:0]    address;
wire [7:0]    out;

reg  [7:0]    rom[255:0];

// ASSIGN STATEMENTS
assign out = rom[address];

// MAIN CODE
initial begin
    rom[0] = 0;
    rom[1] = 0;
    rom[2] = 0;
    rom[3] = 0;
    rom[4] = 0;
```

```
rom[5] = 0;
rom[6] = 0;
rom[7] = 0;
rom[8] = 256;
rom[9] = 256;
rom[10] = 256;
rom[11] = 256;
rom[12] = 256;
rom[13] = 256;
rom[14] = 256;
rom[15] = 256;
rom[16] = 0;
rom[17] = 1;
rom[18] = 2;
rom[19] = 3;
rom[20] = 4;
rom[21] = 5;
rom[22] = 6;
rom[23] = 7;
rom[24] = 248;
rom[25] = 249;
rom[26] = 250;
rom[27] = 251;
rom[28] = 252;
rom[29] = 253;
rom[30] = 254;
rom[31] = 255;
rom[32] = 0;
rom[33] = 2;
rom[34] = 4;
rom[35] = 6;
rom[36] = 8;
rom[37] = 10;
rom[38] = 12;
rom[39] = 14;
rom[40] = 240;
rom[41] = 242;
rom[42] = 244;
rom[43] = 246;
rom[44] = 248;
rom[45] = 250;
rom[46] = 252;
rom[47] = 254;
rom[48] = 0;
rom[49] = 3;
rom[50] = 6;
```

```
rom[51] = 9;
rom[52] = 12;
rom[53] = 15;
rom[54] = 18;
rom[55] = 21;
rom[56] = 232;
rom[57] = 235;
rom[58] = 238;
rom[59] = 241;
rom[60] = 244;
rom[61] = 247;
rom[62] = 250;
rom[63] = 253;
rom[64] = 0;
rom[65] = 4;
rom[66] = 8;
rom[67] = 12;
rom[68] = 16;
rom[69] = 20;
rom[70] = 24;
rom[71] = 28;
rom[72] = 224;
rom[73] = 228;
rom[74] = 232;
rom[75] = 236;
rom[76] = 240;
rom[77] = 244;
rom[78] = 248;
rom[79] = 252;
rom[80] = 0;
rom[81] = 5;
rom[82] = 10;
rom[83] = 15;
rom[84] = 20;
rom[85] = 25;
rom[86] = 30;
rom[87] = 35;
rom[88] = 216;
rom[89] = 221;
rom[90] = 226;
rom[91] = 231;
rom[92] = 236;
rom[93] = 241;
rom[94] = 246;
rom[95] = 251;
rom[96] = 0;
```

```
rom[97] = 6;
rom[98] = 12;
rom[99] = 18;
rom[100] = 24;
rom[101] = 30;
rom[102] = 36;
rom[103] = 42;
rom[104] = 208;
rom[105] = 214;
rom[106] = 220;
rom[107] = 226;
rom[108] = 232;
rom[109] = 238;
rom[110] = 244;
rom[111] = 250;
rom[112] = 0;
rom[113] = 7;
rom[114] = 14;
rom[115] = 21;
rom[116] = 28;
rom[117] = 35;
rom[118] = 42;
rom[119] = 49;
rom[120] = 200;
rom[121] = 207;
rom[122] = 214;
rom[123] = 221;
rom[124] = 228;
rom[125] = 235;
rom[126] = 242;
rom[127] = 249;
rom[128] = 256;
rom[129] = 248;
rom[130] = 240;
rom[131] = 232;
rom[132] = 224;
rom[133] = 216;
rom[134] = 208;
rom[135] = 200;
rom[136] = 64;
rom[137] = 56;
rom[138] = 48;
rom[139] = 40;
rom[140] = 32;
rom[141] = 24;
rom[142] = 16;
```

```
rom[143] = 8;
rom[144] = 256;
rom[145] = 249;
rom[146] = 242;
rom[147] = 235;
rom[148] = 228;
rom[149] = 221;
rom[150] = 214;
rom[151] = 207;
rom[152] = 56;
rom[153] = 49;
rom[154] = 42;
rom[155] = 35;
rom[156] = 28;
rom[157] = 21;
rom[158] = 14;
rom[159] = 7;
rom[160] = 256;
rom[161] = 250;
rom[162] = 244;
rom[163] = 238;
rom[164] = 232;
rom[165] = 226;
rom[166] = 220;
rom[167] = 214;
rom[168] = 48;
rom[169] = 42;
rom[170] = 36;
rom[171] = 30;
rom[172] = 24;
rom[173] = 18;
rom[174] = 12;
rom[175] = 6;
rom[176] = 256;
rom[177] = 251;
rom[178] = 246;
rom[179] = 241;
rom[180] = 236;
rom[181] = 231;
rom[182] = 226;
rom[183] = 221;
rom[184] = 40;
rom[185] = 35;
rom[186] = 30;
rom[187] = 25;
rom[188] = 20;
```

```
rom[189] = 15;
rom[190] = 10;
rom[191] = 5;
rom[192] = 256;
rom[193] = 252;
rom[194] = 248;
rom[195] = 244;
rom[196] = 240;
rom[197] = 236;
rom[198] = 232;
rom[199] = 228;
rom[200] = 32;
rom[201] = 28;
rom[202] = 24;
rom[203] = 20;
rom[204] = 16;
rom[205] = 12;
rom[206] = 8;
rom[207] = 4;
rom[208] = 256;
rom[209] = 253;
rom[210] = 250;
rom[211] = 247;
rom[212] = 244;
rom[213] = 241;
rom[214] = 238;
rom[215] = 235;
rom[216] = 24;
rom[217] = 21;
rom[218] = 18;
rom[219] = 15;
rom[220] = 12;
rom[221] = 9;
rom[222] = 6;
rom[223] = 3;
rom[224] = 256;
rom[225] = 254;
rom[226] = 252;
rom[227] = 250;
rom[228] = 248;
rom[229] = 246;
rom[230] = 244;
rom[231] = 242;
rom[232] = 16;
rom[233] = 14;
rom[234] = 12;
```

```
        rom[235] = 10;
        rom[236] = 8;
        rom[237] = 6;
        rom[238] = 4;
        rom[239] = 2;
        rom[240] = 256;
        rom[241] = 255;
        rom[242] = 254;
        rom[243] = 253;
        rom[244] = 252;
        rom[245] = 251;
        rom[246] = 250;
        rom[247] = 249;
        rom[248] = 8;
        rom[249] = 7;
        rom[250] = 6;
        rom[251] = 5;
        rom[252] = 4;
        rom[253] = 3;
        rom[254] = 2;
        rom[255] = 1;
end
endmodule        // Sm_Rom
```

17.2.3 Partial Sums Algorithm Implementation

The following Verilog code implements signed integer multiplication using what is called the partial sums algorithm, similar to the one used for unsigned integer multiplication. The difference is that when the most significant bit of the multiplier is examined, if it is a 1, the multiplicand is multiplied by -1 rather than one. In other words, the value is subtracted rather than added. This is because the most significant bit of the multiplier is actually a sign bit. A 1 in this position represents a negative number. An example of how this works can be seen in Figures 17-1 and 17-2.

```
              2   1     multiplicand
    x  -  1   0   2     multiplier
          ---------------
              4   2     partial sum
       2  1            partial sum
       --------------------
    -  2  1   4   2     product
```

Figure 17-1 Decimal multiplication of signed integers using partial sums.

```
            0   0   0   1   0   1   0   1   multiplicand
        x   1   0   0   1   1   0   1   0   multiplier
        -------------------------------------------
                    1   0   1   0   1       partial sum
                1   0   1   0   1           partial sum
            1   0   1   0   1               partial sum
        1   0   1   0   1   1               negated sum
    -----------------------------------------------------------
    1   0   1   1   1   1   0   1   0   0   0   1   0   product
```

Figure 17-2 Binary multiplication of signed integers using partial sums.

```
/**********************************************************/
// MODULE:          signed integer multiplier
//
// FILE NAME:       smultiply3_rtl.v
// VERSION:         1.1
// DATE:            January 1, 2003
// AUTHOR:          Bob Zeidman, Zeidman Consulting
//
// CODE TYPE:       Register Transfer Level
//
// DESCRIPTION:  This module defines a multiplier of signed
// integers. This code uses a shift register and partial
// sums.
//
/**********************************************************/

// DEFINES
`define DEL   1      // Clock-to-output delay. Zero
                     // time delays can be confusing
                     // and sometimes cause problems.
`define OP_BITS 4    // Number of bits in each operand
`define OP_SIZE 2    // Number of bits required to
                     // represent the OP_BITS constant
// TOP MODULE
module SignedMultiply(
      clk,
      reset,
      a,
      b,
      multiply_en,
      product,
      valid);
```

```
// PARAMETERS

// INPUTS
input                        clk;              // Clock
input                        reset;            // Reset input
input [`OP_BITS-1:0]          a;                // Multiplicand
input [`OP_BITS-1:0]          b;                // Multiplier
input                        multiply_en;      // Multiply enable input

// OUTPUTS
output [2*`OP_BITS-1:0]       product;          // Product
output                       valid;            // Is output valid yet?

// INOUTS

// SIGNAL DECLARATIONS
wire                         clk;
wire                         reset;
wire  [`OP_BITS-1:0]          a;
wire  [`OP_BITS-1:0]          b;
wire                         multiply_en;
reg   [2*`OP_BITS-1:0]        product;
wire                         valid;

reg   [`OP_SIZE-1:0]          count;            // Count the number
                                               // of bit shifts
reg   [2*`OP_BITS-1:0]        a_ext;            // Sign extended
                                               // version of a

// ASSIGN STATEMENTS
assign valid = (count == 0) ? 1'b1 : 1'b0;

// MAIN CODE

// Look at the rising edge of the clock
// and the rising edge of reset
always @(posedge reset or posedge clk) begin
    if (reset) begin
        count <= #`DEL `OP_SIZE'b0;
    end
    else begin
        if (multiply_en && valid) begin
            // Put the count at all ones - the maximum
            count <= #`DEL ~`OP_SIZE'b0;
        end
        else if (count)
            count <= #`DEL count - 1;
```

```
        end
end

// Look at the rising edge of the clock
always @(posedge clk) begin
    if (multiply_en & valid) begin

        // Create the extended version of a
        a_ext[`OP_BITS-1:0] = a;
        // Sign extend a into a_ext
        if (a[`OP_BITS-1])
            a_ext[2*`OP_BITS-1:`OP_BITS] = ~`OP_BITS'b0;
        else
            a_ext[2*`OP_BITS-1:`OP_BITS] = `OP_BITS'b0;

        if (b[`OP_BITS-1]) begin
            // If the MSB of the multiplier is 1,
            // SUBTRACT the sign extended multiplicand
            // from the product
            product <= #`DEL 0 - a_ext;
        end
        else begin
            // If the MSB of the multiplier is 0,
            // load 0 into the product
            product <= #`DEL 0;
        end
    end
    else if (count) begin
        if (b[count-1]) begin
            // If this bit of the multiplier is 1,
            // shift the product left and add the
            // sign extended multiplicand
            product <= #`DEL (product << 1) + a_ext;
        end
        else begin
            // If this bit of the multiplier is 0,
            // just shift the product left
            product <= #`DEL product << 1;
        end
    end
end
endmodule      // SignedMultiply
```

17.3 SIMULATION CODE

The simulation code generates every possible combination of operands and examines each product for correctness.

```verilog
/*************************************************************/
// MODULE:          signed integer multiplier simulation
//
// FILE NAME:       smultiply_sim.v
// VERSION:         1.1
// DATE:            January 1, 2003
// AUTHOR:          Bob Zeidman, Zeidman Consulting
//
// CODE TYPE:       Simulation
//
// DESCRIPTION:  This module provides stimuli for simulating
// a signed integer multiplier. It generates every possible
// combination of operands and examines each product for
// correctness.
//
/*************************************************************/

// DEFINES
`define OP_BITS 4    // Number of bits in each operand

// TOP MODULE
module smultiply_sim();

// PARAMETERS

// INPUTS

// OUTPUTS

// INOUTS

// SIGNAL DECLARATIONS
reg                     clock;
reg                     reset;
reg   [`OP_BITS-1:0]    a_in;
reg   [`OP_BITS-1:0]    b_in;
reg                     multiply_en;
wire  [2*`OP_BITS-1:0]  product_out;
wire                    valid;

reg   [2*`OP_BITS:0]    cycle_count;   // Counts valid clock
                                       // cycles
integer                 val_count;       // Counts cycles
                                       // between valid data
```

```
integer              a_integer;      // Signed version of
                                     // a_in
integer              b_integer;      // Signed version of
                                     // b_in
reg  [`OP_BITS-1:0]  temp;           // Temporary storage
reg  [2*`OP_BITS-1:0] ltemp;         // Temporary storage
integer              expect_integer; // Expected integer
                                     // product
reg [2*`OP_BITS-1:0] expect;          // Expected output

// ASSIGN STATEMENTS


// MAIN CODE

// Instantiate the multiplier
SignedMultiply smult(
        .clk(clock),
        .reset(reset),
        .a(a_in),
        .b(b_in),
        .multiply_en(multiply_en),
        .product(product_out),
        .valid(valid));

// Initialize inputs
initial begin
    clock = 1;
    cycle_count = 0;

    multiply_en = 1;     // Begin the operation

    reset = 1;           // Toggle reset to initialize
    #10 reset = 0;       // the valid output

    val_count = 0;       // How many cycles to output valid data?
end

// Generate the clock
always #100 clock = ~clock;

// Simulate
always @(negedge clock) begin
    if (valid === 1'b1) begin
        // Check the result for correctness
        if (product_out !== expect) begin
```

```
        $display("\nERROR at time %0t:", $time);
        $display("Adder is not working");
        if (multiply_en)
            $display("Multiplier is enabled");
        else
            $display("Multiplier is not enabled");
        $display("    a_in = %d (%h)", a_integer, a_in);
        $display("    b_in = %d (%h)", b_integer, b_in);
        $display("    expected result = %d (%h)",
                    expect_integer, expect);
        expect_integer = long_to_int(product_out);
        $display("    actual output = %d (%h)\n",
                    expect_integer, product_out);

        // Use $stop for debugging
        $stop;
end


    // Create inputs between 0 and all 1s
    a_in = cycle_count[2*`OP_BITS-1:`OP_BITS];
    b_in = cycle_count[`OP_BITS-1:0];
    // Convert the unsigned numbers to signed numbers
    a_integer = short_to_int(a_in);
    b_integer = short_to_int(b_in);

    // How many cycles to output valid data?
    val_count = 0;

    // Count the valid cycles
    cycle_count = cycle_count + 1;
    if (cycle_count[2*`OP_BITS]) begin
        // We've cycled once
        case (cycle_count[1:0])
            0: begin
                expect_integer = a_integer * b_integer;
                expect = expect_integer[2*`OP_BITS-1:0];
            end
            1: begin
                multiply_en = 0; // Stop the operation

                // We changed the inputs, but don't change the
                // previous expected value since we have
                // stopped the operation
            end
            2: begin
                $display("\nSimulation complete - no errors\n");
                $finish;
```

```
                    end
               endcase
          end
          else begin
               expect_integer = a_integer * b_integer;
               expect = expect_integer[2*`OP_BITS-1:0];
          end
     end
     else begin
          // Keep track of how many cycles to output valid data
          val_count = val_count + 1;

          if (val_count > 2*`OP_BITS+3) begin
               $display("\nERROR at time %0t:", $time);
               $display("Too many cycles for valid data\n");

               // Use $stop for debugging
               $stop;
          end
     end
end

// FUNCTIONS

// Function  to convert a reg of `OP_BITS
// bits to a signed integer
function integer short_to_int;

input [`OP_BITS-1:0]x;

begin
     if (x[`OP_BITS-1]) begin
          // This must be done in two parts so that the
          // simulator does not do any automatic casting
          // on its own.
          temp = ~x + 1;
          short_to_int = 0 - temp;
     end
     else
          short_to_int = x;
end
endfunction

// Function  to convert a reg of 2*`OP_BITS
// bits to a signed integer
function integer long_to_int;
```

```
input [2*`OP_BITS-1:0]  x;

begin
   if (x[2*`OP_BITS-1]) begin
       // This must be done in two parts so that the
       // simulator does not do any automatic casting
       // on its own.
       ltemp = ~x + 1;
       long_to_int = 0 - ltemp;
   end
   else
       long_to_int = x;
end
endfunction
```

ERROR DETECTION AND CORRECTION

*T*he following chapters describe various error detection and correction techniques.

The Parity Generator and Checker

*P*arity generation and parity checking is a simple, but useful form of error detection that is incorporated in many devices. For every n bits on a bus, an extra parity bit is generated and appended to the bus. The two types of parity are even parity and odd parity. If we are using even parity, we count the number of 1's in the data word of n bits. If there are an odd number of 1s, then the parity bit will be a 1 so that, including the parity bit, the number of 1s are an even number. If there are already an even number of bits, then the parity bit is 0. Similarly for odd parity, we want the total number of 1s, including the parity bit, to be an odd number. Some examples of odd and even parity are shown in Table 18-1 below.

The idea behind parity is simple. If there is an error, it will most likely be a single bit error. Parity will flag all single bit errors, although it cannot be used to correct the error. The advantage of using parity is that is requires very little hardware to implement and does not slow down the performance of the system if it is implemented correctly. The disadvantage is that multiple bit errors can go undetected, and it cannot correct errors. Before using parity, it

Table 18-1   Parity Generation

Data word	Even Parity Bit	Odd Parity Bit
00000000	0	1
0101	0	1
1110001100111100	1	0
11111110	1	0

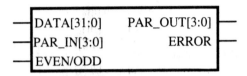

Figure 18-1 Parity generator and checker.

is important to understand what type of errors may occur in your system and what type of error correction, if any, you may need. More sophisticated techniques can be used to correct errors after they have occurred, restoring the original data. More sophisticated error detection and correction hardware is described in later chapters in this section of the book.

The parity generator/checker that is described here is shown in Figure 18-1. It takes in a 32-bit bus plus 4 parity bits and creates the specified parity bits. It also compares the generated parity bits against the expected ones and asserts the ERROR signal if there is a mismatch. Each bit of parity corresponds to a byte of data.

18.1 IMPLEMENTATION CODE

The following code implements a parity generator and checker. Note that due to the simplicity of the circuit, the behavioral and RTL code are the same.

```
/**************************************************************/
// MODULE:          parity generator/checker
//
// FILE NAME:       par.v
// VERSION:         1.1
// DATE:            January 1, 2003
// AUTHOR:          Bob Zeidman, Zeidman Consulting
//
// CODE TYPE:       Behavioral and RTL
//
// DESCRIPTION:  This module defines an parity generator
// and checker. Each bit of parity corresponds to a byte
// of input data.
//
/**************************************************************/
```

```
// DEFINES
`define DEL  1      // Clock-to-output delay. Zero
                    // time delays can be confusing
                    // and sometimes cause problems.

// TOP MODULE
module Parity(
      data,
      par_in,
      even_odd,
      par_out,
      error);

// PARAMETERS

// INPUTS
input [31:0]    data;           // 32-bit data input
input [3:0]     par_in;       // Parity input
input           even_odd;        // Select even or odd parity

// OUTPUTS
output [3:0]    par_out;        // Parity output
output [3:0]    error;          // Error output

// INOUTS

// SIGNAL DECLARATIONS
wire [31:0]     data;
wire [3:0]    par_in;
wire            even_odd;
wire [3:0]    par_out;
wire [3:0]    error;

// ASSIGN STATEMENTS
assign #`DEL par_out[3] = even_odd ? ^data[31:24] :
                  ~^data[31:24];
assign #`DEL par_out[2] = even_odd ? ^data[23:16] :
                  ~^data[23:16];
assign #`DEL par_out[1] = even_odd ? ^data[15:8] :
                  ~^data[15:8];
assign #`DEL par_out[0] = even_odd ? ^data[7:0] :
                  ~^data[7:0];

assign #`DEL error[3] = even_odd ? ^{data[31:24], par_in[3]}:
                  ~^{data[31:24], par_in[3]};
assign #`DEL error[2] = even_odd ? ^{data[23:16], par_in[2]}:
                  ~^{data[23:16], par_in[2]};
assign #`DEL error[1] = even_odd ? ^{data[15:8], par_in[1]} :
```

```
                        ~^{data[15:8], par_in[1]};
assign #`DEL error[0] = even_odd ? ^{data[7:0], par_in[0]} :
                        ~^{data[7:0], par_in[0]};

// MAIN CODE

endmodule     // Parity
```

18.2 SIMULATION CODE

The following code is used to test the parity generation and checking logic. Note that a clock is used in the simulation, even though this is a fully combinatorial circuit. The clock is used simply to allow the simulation to work in phases. In the first phase, the inputs are set up. In the second phase, the outputs are checked for correctness.

Also note that the code used to generate the input parity for simulation is slightly different than the code used in the actual hardware descriptions above. It is always a good idea to generate test code independently of the hardware design code. Otherwise, any incorrect assumption in the hardware design code will be carried into the simulation code. This incorrect assumption may never be discovered through simulation.

```
/*****************************************************/
// MODULE:        parity generator/checker simulation
//
// FILE NAME:     par_sim.v
// VERSION:       1.2
// DATE:          January 1, 2003
// AUTHOR:        Bob Zeidman, Zeidman Consulting
//
// CODE TYPE:     Simulation
//
// DESCRIPTION:  This module provides stimuli for simulating
// a parity generator and checker. It tests a large number of
// random data words using odd parity, then a random number
// of data words using even parity. It then forces errors and
// checks that the bad data words are correctly signaled by
// error output.
//
/*****************************************************/
// DEFINES
`define CYCLE_NUM 100     // Number of cycles per test to
                          // perform
// TOP MODULE
module par_sim();
```

```
// PARAMETERS

// INPUTS

// OUTPUTS

// INOUTS

// SIGNAL DECLARATIONS
reg   [31:0]    data;
reg   [3:0] parity_in;
reg             even_odd;
wire [3:0] parity_out;
wire [3:0] error;

reg             clock;
integer    cycle_count;        // Counter for simulation events
reg             force_error;       // Force an error on the input parity

// ASSIGN STATEMENTS

// MAIN CODE

// Instantiate the parity generator/checker
Parity parity(
        .data(data),
        .par_in(parity_in),
        .even_odd(even_odd),
        .par_out(parity_out),
        .error(error));

// Initialize inputs
initial begin
    cycle_count = 0;
    clock = 1;
    even_odd = 0;          // Look at odd parity first
    force_error = 0;       // Do not force a parity error on inputs
    data = $random(0);         // Initialize random number generator
end

// Generate the clock
always #100 clock = ~clock;

// Simulate
// Set up the inputs on the falling edge of the clock
always @(negedge clock) begin
```

```verilog
    case (cycle_count)
       `CYCLE_NUM:   begin
          // Switch the even/odd input
          even_odd = ~even_odd;
       end
       2*`CYCLE_NUM: begin
          // Force an error to check that the error
          // output is asserted
          force_error = 1;
       end
       3*`CYCLE_NUM: begin
          $display("\nSimulation complete - no errors\n");
          $finish;
       end
    endcase

    data = {$random} % 32'hFFFFFFFF;
    if (even_odd ^ force_error) begin
        parity_in[3] = ^data[31:24];
        parity_in[2] = ^data[23:16];
        parity_in[1] = ^data[15:8];
        parity_in[0] = ^data[7:0];
    end
    else begin
        parity_in[3] = ~^data[31:24];
        parity_in[2] = ~^data[23:16];
        parity_in[1] = ~^data[15:8];
        parity_in[0] = ~^data[7:0];
    end
end

// Check the outputs on the rising edge of the clock
always @(posedge clock) begin
    if (~force_error) begin
        if (parity_in !== parity_out) begin
            $display("\nERROR at time %0t:", $time);
            $display("Output parity is incorrect");
            $display("    expected = %h", parity_in);
            $display("    actual   = %h\n", parity_out);

            // Use $stop for debugging
            $stop;
        end
        if (error !== 0) begin
            $display("\nERROR at time %0t:", $time);
            $display("Error signal is asserted");
            $display("    error = %b\n", error);
```

```verilog
            // Use $stop for debugging
            $stop;
        end
    end
    else begin
        if (parity_in !== ~parity_out) begin
            $display("\nERROR at time %0t:", $time);
            $display("Output parity is incorrect");
            $display("    expected = %h", ~parity_in);
            $display("    actual   = %h\n", parity_out);

            // Use $stop for debugging
            $stop;
        end
        if (error === 0) begin
            $display("\nERROR at time %0t:", $time);
            $display("Error signal is not asserted\n");

            // Use $stop for debugging
            $stop;
        end
    end

    // Increment the cycle count
    cycle_count = cycle_count + 1;
end

endmodule    // par_sim
```

Hamming Code Logic

*H*amming codes are based on the work of R.W. Hamming at Bell Laboratories in 1950. Hamming discovered that by taking a data word and adding bits to it according to an algorithm, errors that cause bits to be changed can be detected and even corrected to obtain the original data. This is a very powerful concept. Parity, which is the most common form of error detection, can only report that an error occurred. Hamming codes have the advantage that they can actually correct the error. It is my personal feeling that Hamming codes would be more widely used if more engineers knew how to implement them. The algorithm is a little involved, but it turns out that the circuitry needed to implement the algorithm is very simple combinatorial logic. In this section, I will introduce the basic theory and then show implementations of the algorithm. The Verilog code for generating and decoding Hamming codes is slightly different for each word size. For this reason, I have included C code for generating the Verilog code. There are two C programs; one is for producing Verilog code to generate Hamming codes and one is for producing Verilog code to decode Hamming codes. Both programs take the word size as an input parameter and generate the appropriate Verilog function.

The basic concept of Hamming codes relies on the concept of Hamming distances:

Hamming Distance: The Hamming distance between two binary numbers is the total number of bit positions in which they differ.

For example, the Hamming distance between the numbers 01110001 and 11010011 is 3 because there are three places where the bits are different. If we were to transmit data that represented a YES or NO, we could represent YES as 1 and NO as 0. If the bit gets corrupted during transmission, a valid YES becomes a valid NO and vice versa. This is a dangerous situation, especially if the question was something like, "Shall we launch our nuclear weapons, Mr. President?" The reason for this problem is that the Hamming distance between the two codes is only 1. If we use the codes of 01 to represent YES and 10 to represent NO, a single bit error will result in an invalid code and will always be detected. The Hamming distance between the two codes is 2. In this case, we can detect all single bit errors, but we cannot correct them.

An even better code would be to use 001 for YES and 110 for NO. Here the Hamming distance between the codes is 3. Because it is an odd number, if there is a single bit error, we can make a best guess as to what the correct message should be by taking the valid answer that is closest (i.e., has the smallest Hamming distance to) the received code. If we receive a 011, we can assume that there was a single bit error, and that our answer is really YES since 011 has a Hamming distance of 1 to the valid code 001 and a Hamming distance of 2 to 110. Of course, in the case of nuclear war, we may still want to ask this question again, rather than assume there was a single bit error.

If we use a code in which each answer has a Hamming distance of 4, we can correct all 1-bit errors and detect, but not correct all 2-bit errors. This is because a 2-bit error ends up with a code that is equal distance between two valid codes. The general formula for a Hamming code that can correct c errors and detect an additional d errors is:

$$2c + d + 1 \leq H_d$$

where H_d is the minimum distance of every valid code to every other valid code.

So in order to transmit data and detect and correct errors, we must map every data word to a set of Hamming codes. We can do this by appending bits to each data word to increase their Hamming distances. The next questions are how to determine the number of bits to append and what their values need to be. I won't go into all of the theory here, but I will explain the practical approach. First, we create the smallest matrix H that has the following properties.

Matrix Property 1: The number of columns of matrix H is $n + m$, where n is the number of code bits and m is the number of data bits. Find the smallest n such that

$$n + m \leq 2^n - 1$$

Matrix Property 2: The number of rows of matrix H is n. This is also the number of code bits to be appended.**Matrix Property 3:** The left side of matrix H is an $n \times n$ identity matrix.

Matrix Property 4: The remaining columns of matrix H are created such that each column is a unique binary number.

Matrix Property 5: Once we have created matrix H, we use it to find the Hamming code bits for every data word using the following equation:

$$H \bullet x^{\mathsf{T}} = 0$$

where x is the data word with the Hamming code bits appended to the most significant bits. The T superscript simply means that the vector is "transposed" or turned on its side in order to do the matrix multiplication. Also, the 0 is really an n-bit vector of all zeros.

For a 4-bit data word, we would create the following 3×7 matrix:

$$H = \begin{array}{ccccccc} 1 & 0 & 0 & 1 & 1 & 0 & 1 \\ 0 & 1 & 0 & 1 & 0 & 1 & 1 \\ 0 & 0 & 1 & 0 & 1 & 1 & 1 \end{array}$$

We append the currently unknown code bits onto the data word to create a vector 7-bit vector:

$$x = \quad c_2 \quad c_1 \quad c_0 \quad d_3 \quad d_2 \quad d_1 \quad d_0$$

where c_i are the code bits and d_i are the data bits. Using the matrix equation of property 5, we end up with 3 equations for 3 code bits, which can easily be solved using combinatorial logic. Remember that single bit addition and single bit subtraction are both the same as the XOR function. Given this fact, we end up with

$$c_2 = d_3 \wedge d_2 \wedge d_0$$

$$c_1 = d_3 \wedge d_1 \wedge d_0$$

$$c_0 = d_2 \wedge d_1 \wedge d_0$$

Now that we have determined how to create the code bits, we need to know how to take data that we receive and correct the errors to produce the original data. To do this, we create a syndrome according to the following rules:**Syndrome Rule 1:** Use the following equation to produce a syndrome s.

$$H \bullet x'^{\mathsf{T}} = s$$

where x' is the data plus code bits that were received.

Syndrome Rule 2: If the syndrome is all zeros, no single-bit error has occurred. In this case, $x' = x$. We simply strip off the code bits to obtain the original data.

Syndrome Rule 3: If the syndrome is not all zeros, the column in the matrix that matches the transposed syndrome corresponds to the bit that is in error. Change that bit in the received data and then strip off the code bits to obtain the correct data. For example, if the transposed syndrome equals column 3 of the matrix, flip bit 3 of the received data. Of course if it was one of the code bits that got corrupted, no action needs to be taken before stripping of the code bits.

In our example above, using a 4-bit data word, suppose we transmit the following data plus code bits.

$$x = \quad 1 \quad 1 \quad 0 \quad 0 \quad 1 \quad 1 \quad 0$$

Now suppose that the data gets corrupted in transmission and is received as follows.

$$x' = \quad 1 \quad 1 \quad 0 \quad 0 \quad 1 \quad 0 \quad 0$$

Multiplying x' by the matrix H, we produce the following syndrome.

$$s = \quad 1 \quad 1 \quad 0$$

This syndrome corresponds to column 1 of matrix H (counting from 0 at the far right, just like counting bits in a data word). This tells us that bit 1 is the corrupted bit, which is correct.

Now that all of the theory has been discussed, here is the actual Verilog code for implementing Hamming codes for data of any width.

19.1 IMPLEMENTATION CODE

Because the logic for generating and decoding Hamming codes is entirely combinatorial, the Behavioral Level code is identical to the RTL code.

19.1.1 C Code for Hamming Code Generation

The following C code creates a program that generates the Verilog code for generating Hamming codes. The program is run by entering:

```
hamgen n filename
```
where n is the width of the data word in bits, and *filename* is the name of the Verilog file output.

```
/*************************************************************/
/*************************************************************/
/*****                      HAMGEN                     *****/
```

```
/***** file: hamgen.c                                 *****/
/*****                                                *****/
/***** Written by Bob Zeidman                         *****/
/***** Zeidman Consulting                             *****/
/*****                                                *****/
/***** Description:                                   *****/
/***** This program generates a Verilog module for    *****/
/***** generating single-bit error detection and      *****/
/***** correction bits of a data word with an         *****/
/***** arbitrary number of bits                       *****/
/*****                                                *****/
/***** Version   Date      Author       Comments      *****/
/***** 1.0       8/25/98   Bob Zeidman   Original program *****/
/***** 1.1       3/26/01   Bob Zeidman   Fix problem found*****/
/*****           by David Morton where hex subscripts   *****/
/*****           were used instead of decimal subscripts *****/
/*****           for some signals                     *****/
/***** 1.2       4/1/03 Bob Zeidman   Changed placement *****/
/*****           of parameters in Verilog code output  *****/
/*****                                                *****/
/********************************************************/
/********************************************************/


/***** Library Include Files *****/
/* ANSI C libraries */
#include <stdlib.h>          /* standard library */
#include <stdio.h>           /* standard io library */
#include <string.h>          /* string library */
#include <time.h>            /* time library */


/***** Global variables *****/


/***** Constants *****/
#define VERSION "1.1"        /* Program version */
#define FILELEN 76           /* Maximum number of characters
                                in a file name */


/***** SUBROUTINES *****/


/***** Describe the usage of the program for the user *****/
void describe()
{
    printf("\nError Detecting and Correcting Code ");
    printf("Generator Version %s\n", VERSION);
    printf("This program creates a Verilog module that ");
    printf("generates Hamming codes");
    printf("for a data word of any width.\n");
    printf("\nUsage:\n");
    printf("\n\thamgen width filename\n\n");
    printf("where width is the bit width of the ");
    printf("data word.\n");
```

```
   printf("       filename is the name of the file to ");
   printf("generate.\n");
}
/***** MAIN ROUTINE *****/
/***** argc        Argument count        *****/
/***** argv[] Argument variables   *****/
void main(int argc, char *argv[])
{
   int width;                  /* width of operands in bits */
   int edc_bits;               /* number of edc bits to append */
   int total_bits;             /* number of data plus edc bits */
   int *matrix;                /* matrix for edc generation */
   int *syndrome;              /* list of syndromes */
   char filename[FILELEN];     /* Verilog file name */
   FILE *VerilogFile;          /* Verilog output file */
   time_t filetime;            /* Output file creation time */
   int start;                  /* Have we started creating
                                  the assign statements? */
   int i;                      /* all-purpose integer variable */
   int j;                      /* all-purpose integer variable */

   /* Check for exactly two arguments */
   if (argc != 3)
   {
      /* if there's not two arguments, it's wrong */
      /* describe the correct usage of this program,
         then exit */
      describe();
      exit(1);
   }

   /* Convert the first argument to an integer width */
   width = atoi(argv[1]);
   if (width <= 0)
   {
      /* if the width is not a whole number, it's wrong */
      /* describe the correct usage of this program,
         then exit */
      describe();
      exit(1);
   }

   /* Get the filename */
   strncpy(filename, argv[2], FILELEN-1);

   /* Open the file for writing */
   VerilogFile = fopen(filename, "wt");
   if (VerilogFile == NULL)
   {
      printf("Error: Cannot open file %s for writing!\n",
         filename);
```

```
      exit(1);
}

/* Determine the number of total EDC bits to append */
edc_bits = 1;
while ((1 << edc_bits) - edc_bits - 1 < width)
    edc_bits++;

/* Calculate the total number of bits */
total_bits = edc_bits + width;

/* Allocate memory for the code generation matrix */
matrix = malloc(total_bits * sizeof(int));
if (matrix == NULL)
{
    printf("Error: Not enough memory!\n");
    exit(1);
}

/* Allocate memory for the syndrome list */
syndrome = malloc((total_bits+1) * sizeof(int));
if (syndrome == NULL)
{
    printf("Error: Not enough memory!\n");
    free(matrix);
    exit(1);
}

/* Initialize the syndrome list */
for (i = 0; i < total_bits+1; i++)
    syndrome[i] = 0;

j = 1;
for (i = 0; i < total_bits; i++)
{
    matrix[i] = j;
    syndrome[j] = 1;
    if (i < edc_bits-1)
        j = j << 1;
    else if (i == edc_bits-1)
        j = 3;
    else if (i < total_bits-1)
    {
        do
        {
            j++;
        } while (syndrome[j]);
    }
}
```

```
}
/* Make the file creation date string */
time (&filetime);

fprintf(VerilogFile, "/****************************");
fprintf(VerilogFile, "*************************/\n");
fprintf(VerilogFile, "// MODULE:          Hamming Code ");
fprintf(VerilogFile, "Generator\n");
fprintf(VerilogFile, "//                  for %d-bit ",
    width);
fprintf(VerilogFile, "data\n");
fprintf(VerilogFile, "//\n");
fprintf(VerilogFile, "// FILE NAME:       %s\n", filename);
fprintf(VerilogFile, "// VERSION:         %s\n", VERSION);
fprintf(VerilogFile, "// DATE:            %s",
    ctime(&filetime));
fprintf(VerilogFile, "// AUTHOR:          HAMGEN.EXE\n");
fprintf(VerilogFile, "//\n");
fprintf(VerilogFile, "// CODE TYPE:");
fprintf(VerilogFile, "      Register Transfer Level\n");
fprintf(VerilogFile, "//\n");
fprintf(VerilogFile, "// DESCRIPTION:    This module ");
fprintf(VerilogFile, "defines a generator of codes\n");
fprintf(VerilogFile, "// for detecting and correcting");
fprintf(VerilogFile, " single bit data errors.\n");
fprintf(VerilogFile, "//\n");
fprintf(VerilogFile, "/****************************");
fprintf(VerilogFile, "*************************/\n");
fprintf(VerilogFile, "\n");
fprintf(VerilogFile, "// DEFINES\n");
fprintf(VerilogFile, "`define DEL 1        ");
fprintf(VerilogFile, "// Clock-to-output delay. Zero\n");
fprintf(VerilogFile, "                          ");
fprintf(VerilogFile, "// time delays can be confusing\n");
fprintf(VerilogFile, "                          ");
fprintf(VerilogFile, "// and sometimes cause ");
fprintf(VerilogFile, "problems.\n");
fprintf(VerilogFile, "\n");
fprintf(VerilogFile, "// TOP MODULE\n");
fprintf(VerilogFile, "module HamGen(\n");
fprintf(VerilogFile, "        data_in,\n");
fprintf(VerilogFile, "        edc_out);\n");
fprintf(VerilogFile, "\n");
fprintf(VerilogFile, "// PARAMETERS\n");
fprintf(VerilogFile, "\n");
fprintf(VerilogFile, "// INPUTS\n");
fprintf(VerilogFile, "input [%d:0]    data_in", width-1);
fprintf(VerilogFile, ";    // Input data\n");
fprintf(VerilogFile, "\n");
```

```c
    fprintf(VerilogFile, "// OUTPUTS\n");
    fprintf(VerilogFile, "output [%d:0]     edc_out; ",
        edc_bits-1);
    fprintf(VerilogFile, "\n");
    fprintf(VerilogFile, "// INOUTS\n");
    fprintf(VerilogFile, "\n");
    fprintf(VerilogFile, "// SIGNAL DECLARATIONS\n");
    fprintf(VerilogFile, "wire [%d:0]     data_in;\n",
        width-1);
    fprintf(VerilogFile, "wire [%d:0]     edc_out;\n",
        edc_bits-1);
    fprintf(VerilogFile, "\n");
    fprintf(VerilogFile, "// ASSIGN STATEMENTS\n");
    /* Create the combinatorial logic to generate edc bits */
    for (i = 0; i < edc_bits; i++)
    {
        start = 0;
        fprintf(VerilogFile, "assign #`DEL edc_out[%x] = ",
            edc_bits-i-1);
        for (j = edc_bits; j < total_bits; j++)
        {
            if (matrix[j] & (1 << i))
            {
                if (start)
                    fprintf(VerilogFile, " ^ ");
                fprintf(VerilogFile, "data_in[%x]",
                    total_bits-j-1);
                start = 1;
            }
        }
        fprintf(VerilogFile, ";\n");
    }

    fprintf(VerilogFile, "\n");
    fprintf(VerilogFile, "// MAIN CODE\n");
    fprintf(VerilogFile, "\n");
    fprintf(VerilogFile, "endmodule        // HamGen\n");

    /* Free up the allocated memory */
    free(matrix);
    free(syndrome);
}
```

19.1.2 Verilog Code for Hamming Code Generation

The following Verilog code generates Hamming codes for 8-bit data words. It was
generated using the C program above.

```
/**************************************************************/
// MODULE:          Hamming Code Generator
//                  for 8-bit data
//
// FILE NAME:       hamgen.v
// VERSION:         1.1
// DATE:            Mon Apr 07 07:42:31 2003
// AUTHOR:          HAMGEN.EXE
//
// CODE TYPE:       Register Transfer Level
//
// DESCRIPTION:     This module defines a generator of codes
// for detecting and correcting single bit data errors.
//
/**************************************************************/

// DEFINES
`define DEL   1          // Clock-to-output delay. Zero
                         // time delays can be confusing
                         // and sometimes cause problems.
// TOP MODULE
module HamGen(
        data_in,
        edc_out);

// PARAMETERS

// INPUTS
input [7:0]      data_in;    // Input data

// OUTPUTS
output [3:0]     edc_out;    // EDC output
// INOUTS

// SIGNAL DECLARATIONS
wire [7:0]       data_in;
wire [3:0]       edc_out;

// ASSIGN STATEMENTS
assign #`DEL edc_out[3] = data_in[7] ^ data_in[6] ^
    data_in[4] ^ data_in[3] ^ data_in[1];
assign #`DEL edc_out[2] = data_in[7] ^ data_in[5] ^
```

```
    data_in[4] ^ data_in[2] ^ data_in[1];
assign #`DEL edc_out[1] = data_in[6] ^ data_in[5] ^
    data_in[4] ^ data_in[0];
assign #`DEL edc_out[0] = data_in[3] ^ data_in[2] ^
    data_in[1] ^ data_in[0];

// MAIN CODE

endmodule        // HamGen
```

19.1.3 C Code for Hamming Code Decoding

The following C code creates a program that generates the Verilog code for decoding
Hamming codes. The program is run by entering:

```
        hamdec n filename
```

where *n* is the width of the data word in bits, and *filename* is the name of the Verilog file
output.

```
/***********************************************************/
/***********************************************************/
/*****                      HAMDEC                    *****/
/***** file: hamdec.c                                 *****/
/*****                                                *****/
/***** Written by Bob Zeidman                         *****/
/***** Zeidman Consulting                             *****/
/*****                                                *****/
/***** Description:                                   *****/
/***** This program generates a Verilog module for    *****/
/***** detecting and correcting single-bit errors in  *****/
/***** data of an arbitrary bit width                 *****/
/*****                                                *****/
/***** Version   Date      Author        Comments     *****/
/***** 1.0       8/26/98   Bob Zeidman   Original program *****/
/***** 1.1       3/26/01   Bob Zeidman   Fix problem found*****/
/*****           by David Morton where hex subscripts  *****/
/*****           were used instead of decimal subscripts *****/
/*****           for some signals                      *****/
/***** 1.2       4/1/03 Bob Zeidman   Changed placement *****/
/*****           of parameters in Verilog code output  *****/
/*****                                                *****/
/***********************************************************/
/***********************************************************/
```

```
/***** Library Include Files *****/
/* ANSI C libraries */
#include <stdlib.h>              /* standard library */
#include <stdio.h>               /* standard io library */
#include <string.h>              /* string library */
#include <time.h>                /* time library */
/***** Global variables *****/

/***** Constants *****/
#define VERSION "1.1"            /* Program version */
#define FILELEN 76               /* Maximum number of characters
                                    in a file name */

/***** SUBROUTINES *****/

/***** Describe the usage of the program for the user *****/
void describe()
{
    printf("\nError Detector and Corrector Generator ");
    printf("Version %s\n", VERSION);
    printf("This program creates a Verilog module that ");
    printf("performs error detecting and correcting ");
    printf("for a data word of any width.\n");
    printf("\nUsage:\n");
    printf("\n\tedcdec width filename\n\n");
    printf("where width is the bit width of the ");
    printf("data word.\n");
    printf("       filename is the name of the file to ");
    printf("generate.\n");
}

/***** MAIN ROUTINE *****/
/***** argc        Argument count       *****/
/***** argv[] Argument variables  *****/
void main(int argc, char *argv[])
{
    int width;                   /* width of operands in bits */
    int edc_bits;                /* number of edc bits to append */
    int total_bits;              /* number of data plus edc bits */
    int *matrix;                 /* matrix for edc generation */
    int *syndrome;               /* list of syndromes */
    char filename[FILELEN];      /* Verilog file name */
    FILE *VerilogFile;           /* Verilog output file */
    time_t filetime;             /* Output file creation time */
    int start;                   /* Have we started creating
                                    the assign statements? */
    int i;                       /* all-purpose integer variable */
    int j;                       /* all-purpose integer variable */

    /* Check for exactly two arguments */
    if (argc != 3)
    {
```

```
        /* if there's not two arguments, it's wrong */
        /* describe the correct usage of this program,
            then exit */
        describe();
        exit(1);
    }

    /* Convert the first argument to an integer width */
    width = atoi(argv[1]);
    if (width <= 0)
    {
        /* if the width is not a whole number, it's wrong */
        /* describe the correct usage of this program,
            then exit */
        describe();
        exit(1);
    }

    /* Get the filename */
    strncpy(filename, argv[2], FILELEN-1);

    /* Open the file for writing */
    VerilogFile = fopen(filename, "wt");
    if (VerilogFile == NULL)
    {
        printf("Error: Cannot open file %s for writing!\n",
            filename);
        exit(1);
    }

    /* Determine the number of EDC bits to append */
    edc_bits = 1;
    while ((1 << edc_bits) - edc_bits - 1 < width)
        edc_bits++;

    /* Calculate the total number of bits */
    total_bits = edc_bits + width;

    /* Allocate memory for the code generation matrix */
    matrix = malloc(total_bits * sizeof(int));
    if (matrix == NULL)
    {
        printf("Error: Not enough memory!\n");
        exit(1);
    }

    /* Allocate memory for the syndrome list */
    syndrome = malloc((total_bits+1) * sizeof(int));
    if (syndrome == NULL)
    {
        printf("Error: Not enough memory!\n");
```

```
    free(matrix);
    exit(1);
}

/* Initialize the syndrome list */
for (i = 0; i < total_bits+1; i++)
    syndrome[i] = 0;

j = 1;
for (i = 0; i < total_bits; i++)
{
    matrix[i] = j;
    syndrome[j] = 1;
    if (i < edc_bits-1)
        j = j << 1;
    else if (i == edc_bits-1)
        j = 3;
    else if (i < total_bits-1)
    {
        do
        {
            j++;
        } while (syndrome[j]);
    }
}
/* Make the file creation date string */
time (&filetime);

fprintf(VerilogFile, "/****************************");
fprintf(VerilogFile, "***************************/\n");
fprintf(VerilogFile, "// MODULE:          ");
fprintf(VerilogFile, "Hamming Code Decoder\n");
fprintf(VerilogFile, "//                  for %d-bit ",
    width);
fprintf(VerilogFile, "data\n");
fprintf(VerilogFile, "//\n");
fprintf(VerilogFile, "// FILE NAME:       %s\n", filename);
fprintf(VerilogFile, "// VERSION:         %s\n", VERSION);
fprintf(VerilogFile, "// DATE:            %s",
    ctime(&filetime));
fprintf(VerilogFile, "// AUTHOR:          HAMDEC.EXE\n");
fprintf(VerilogFile, "//\n");
fprintf(VerilogFile, "// CODE TYPE:");
fprintf(VerilogFile, "       Register Transfer Level\n");
fprintf(VerilogFile, "//\n");
fprintf(VerilogFile, "// DESCRIPTION:    This module ");
fprintf(VerilogFile, "defines an error detector and\n");
```

```
fprintf(VerilogFile, "// corrector of single bit errors");
fprintf(VerilogFile, " using Hamming codes.\n");
fprintf(VerilogFile, "//\n");
fprintf(VerilogFile, "/****************************");
fprintf(VerilogFile, "****************************/\n");
fprintf(VerilogFile, "\n");
fprintf(VerilogFile, "// DEFINES\n");
fprintf(VerilogFile, "`define DEL 1      ");
fprintf(VerilogFile, "// Clock-to-output delay. Zero\n");
fprintf(VerilogFile, "                        ");
fprintf(VerilogFile, "// time delays can be confusing\n");
fprintf(VerilogFile, "                        ");
fprintf(VerilogFile, "// and sometimes cause ");
fprintf(VerilogFile, "problems.\n");
fprintf(VerilogFile, "\n");
fprintf(VerilogFile, "// TOP MODULE\n");
fprintf(VerilogFile, "module HamDec(\n");
fprintf(VerilogFile, "        data_in,\n");
fprintf(VerilogFile, "        edc_in,\n");
fprintf(VerilogFile, "        data_out,\n");
fprintf(VerilogFile, "        error);\n");
fprintf(VerilogFile, "\n");
fprintf(VerilogFile, "// PARAMETERS\n");
fprintf(VerilogFile, "\n");
fprintf(VerilogFile, "// INPUTS\n");
fprintf(VerilogFile, "input [%d:0]  data_in", width-1);
fprintf(VerilogFile, ";        // Input data\n");
fprintf(VerilogFile, "input [%d:0]  edc_in", edc_bits-1);
fprintf(VerilogFile, ";         // EDC bits\n");
fprintf(VerilogFile, "\n");
fprintf(VerilogFile, "// OUTPUTS\n");
fprintf(VerilogFile, "output [%d:0]  data_out; ",
    width-1);
fprintf(VerilogFile, "     // data output\n");
fprintf(VerilogFile, "output          error; ");
fprintf(VerilogFile, "          // Did an error occur?\n");
fprintf(VerilogFile, "\n");
fprintf(VerilogFile, "// INOUTS\n");
fprintf(VerilogFile, "\n");
fprintf(VerilogFile, "// SIGNAL DECLARATIONS\n");
fprintf(VerilogFile, "wire [%d:0]   data_in;\n",
    width-1);
fprintf(VerilogFile, "wire [%d:0]    edc_in;\n",
    edc_bits-1);
fprintf(VerilogFile, "reg  [%d:0]    data_out;\n",
    width-1);
fprintf(VerilogFile, "reg           error;\n");
fprintf(VerilogFile, "\n");
```

```c
fprintf(VerilogFile, "wire [%d:0]     syndrome;\n",
    edc_bits-1);
fprintf(VerilogFile, "\n");
fprintf(VerilogFile, "// ASSIGN STATEMENTS\n");


/* Create the combinatorial logic for the syndrome bits */
for (i = 0; i < edc_bits; i++)
{
    start = 0;
    fprintf(VerilogFile, "assign #`DEL syndrome[%x] = ",
        i);
    for (j = 0; j < total_bits; j++)
    {
        if (matrix[j] & (1 << i))
        {
            if (start)
                fprintf(VerilogFile, " ^ ");
            if (j < edc_bits)
                fprintf(VerilogFile, "edc_in[%x]",
                    edc_bits-j-1);
            else
                fprintf(VerilogFile, "data_in[%x]",
                    total_bits-j-1);
            start = 1;
        }
    }
    fprintf(VerilogFile, ";\n");
}

fprintf(VerilogFile, "\n");
fprintf(VerilogFile, "// MAIN CODE\n");
fprintf(VerilogFile, "\n");
fprintf(VerilogFile, "always @(syndrome or data_in) ");
fprintf(VerilogFile, " begin\n");
fprintf(VerilogFile, "    data_out = data_in;\n");
fprintf(VerilogFile, "\n");
fprintf(VerilogFile, "    case (syndrome)      ");
fprintf(VerilogFile, "// synthesis ");
fprintf(VerilogFile, "parallel_case full_case\n");

/* Generate the case statements */
fprintf(VerilogFile, "        %d'h0: begin\n", edc_bits);
fprintf(VerilogFile, "            error = 0;\n");
fprintf(VerilogFile, "        end\n");
```

```
    for (j = 0; j < total_bits; j++)
    {
        fprintf(VerilogFile, "         %d'h%x: begin\n",
            edc_bits, matrix[j]);
        if (j >= total_bits-width)
        {
            fprintf(VerilogFile, "             data_out[%d] = ",
                total_bits-j-1);
            fprintf(VerilogFile, "~data_in[%d];\n",
                total_bits-j-1);
        }
        fprintf(VerilogFile, "             error = 1;\n");
        fprintf(VerilogFile, "         end\n");
    }

    fprintf(VerilogFile, "    endcase\n");
    fprintf(VerilogFile, "end\n");
    fprintf(VerilogFile, "endmodule        // HamDec\n");

    /* Free up the allocated memory */
    free(matrix);
    free(syndrome);
}
```

19.1.4 Verilog Code for Hamming Code Decoding

The following Verilog code decodes Hamming codes for 8-bit data words. It was generated using the C program above.

```
/**********************************************************/
// MODULE:        Hamming Code Decoder
//                for 8-bit data
//
// FILE NAME:     hamdec.v
// VERSION:       1.1
// DATE:          Mon Apr 07 07:47:10 2003
// AUTHOR:        HAMDEC.EXE
//
// CODE TYPE:     Register Transfer Level
//
// DESCRIPTION:   This module defines an error detector and
// corrector of single bit errors using Hamming codes.
//
/**********************************************************/
```

```
// DEFINES
`define DEL   1           // Clock-to-output delay. Zero
                         // time delays can be confusing
                         // and sometimes cause problems.

// TOP MODULE
module HamDec(
        data_in,
        edc_in,
        data_out,
        error);

// PARAMETERS

// INPUTS
input [7:0]    data_in;      // Input data
input [3:0]    edc_in;       // EDC bits

// OUTPUTS
output [7:0]   data_out;     // data output
output         error;        // Did an error occur?

// INOUTS

// SIGNAL DECLARATIONS
wire [7:0]     data_in;
wire [3:0]     edc_in;
reg  [7:0]     data_out;
reg            error;

wire [3:0]     syndrome;

// ASSIGN STATEMENTS
assign #`DEL syndrome[0] = edc_in[3] ^ data_in[7] ^
    data_in[6] ^ data_in[4] ^ data_in[3] ^ data_in[1];
assign #`DEL syndrome[1] = edc_in[2] ^ data_in[7] ^
    data_in[5] ^ data_in[4] ^ data_in[2] ^ data_in[1];
assign #`DEL syndrome[2] = edc_in[1] ^ data_in[6] ^
    data_in[5] ^ data_in[4] ^ data_in[0];
assign #`DEL syndrome[3] = edc_in[0] ^ data_in[3] ^
    data_in[2] ^ data_in[1] ^ data_in[0];

// MAIN CODE

always @(syndrome or data_in) begin
    data_out = data_in;
```

```
case (syndrome)      // synthesis parallel_case full_case
    4'h0: begin
        error = 0;
    end
    4'h1: begin
        error = 1;
    end
    4'h2: begin
        error = 1;
    end
    4'h4: begin
        error = 1;
    end
    4'h8: begin
        error = 1;
    end
    4'h3: begin
        data_out[7] = ~data_in[7];
        error = 1;
    end
    4'h5: begin
        data_out[6] = ~data_in[6];
        error = 1;
    end
    4'h6: begin
        data_out[5] = ~data_in[5];
        error = 1;
    end
    4'h7: begin
        data_out[4] = ~data_in[4];
        error = 1;
    end
    4'h9: begin
        data_out[3] = ~data_in[3];
        error = 1;
    end
    4'ha: begin
        data_out[2] = ~data_in[2];
        error = 1;
    end
    4'hb: begin
        data_out[1] = ~data_in[1];
        error = 1;
    end
    4'hc: begin
        data_out[0] = ~data_in[0];
        error = 1;
    end
```

```
    endcase
end
endmodule       // HamDec
```

19.2 SIMULATION CODE

A sequence of data words is generated using a simple counter. The data word is put through a Hamming code generator and the error detecting and correcting bits are appended to the word. Then for the first word, no bits are corrupted. For the second word, bit 1 is corrupted. For the third word, bit 2 is corrupted. When the most significant bit is corrupted, the sequence begins again with no bits corrupted. Each time, the data is sent through the Hamming code decoder to obtain the correct data. This data is compared to the original data, which should match perfectly or an error message is displayed and the simulation stops. The simulation normally increments the data word until it wraps back around to zero. If this happens and no errors were found, the simulation ends successfully.

```
/**********************************************************/
// MODULE:       Hamming Code simulation
//
// FILE NAME:    ham_sim.v
// VERSION:      1.1
// DATE:         January 1, 2003
// AUTHOR:       Bob Zeidman, Zeidman Consulting
//
// CODE TYPE:    Simulation
//
// DESCRIPTION:  This module provides stimuli for simulating
// single bit error detecting and correcting logic. A
// sequence of data words is generated using a counter. The
// Hamming code bits are appended to the word. For the first
// word, no bits are corrupted. For the second word, bit 1 is
// corrupted, etc. When the most significant bit is
// corrupted, the sequence begins again with no bits
// corrupted. Each time, the data is sent through decoder to
// obtain the correct data. This data is compared to the
// original data, which should match perfectly or an error
// message is displayed and the simulation stops. The
// simulation increments the data word until it wraps back
// around to zero. If this happens and no errors were found,
// the simulation ends successfully.
//
/**********************************************************/
```

```
// DEFINES
`define DEL   1          // Clock-to-output delay. Zero
                         // time delays can be confusing
                         // and sometimes cause problems.
`define BITS 8           // Number of bits in a data word
`define EDC_BITS 4       // Number of EDC bits

// TOP MODULE
module ham_sim();

// PARAMETERS

// INPUTS

// OUTPUTS

// INOUTS

// SIGNAL DECLARATIONS
reg  [`BITS-1:0]          data_in;
wire [`EDC_BITS-1:0]        edc;
wire [`BITS-1:0]          data_out;
wire                      error;

wire [`BITS-1:0]          new_data;    // New data,
                                       // possibly corrupted
wire [`EDC_BITS-1:0]        new_edc;     // New EDC bits,
                                       // possibly corrupted
reg                       clock;       // Clock
reg  [`EDC_BITS+`BITS-1:0] bit_err;      // Used to force a
                                       // single bit error
reg                       start;       // Have we started?

// ASSIGN STATEMENTS
assign #`DEL new_data = data_in ^ bit_err[`BITS-1:0];
assign #`DEL new_edc =edc ^ bit_err[`EDC_BITS+`BITS-1:`BITS];

// MAIN CODE

// Instantiate the EDC generator
HamGen hamgen(
        .data_in(data_in),
        .edc_out(edc));

// Instantiate the EDC decoder
HamDec hamdec(
      .data_in(new_data),
```

```
               .edc_in(new_edc),
               .data_out(data_out),
               .error(error));

// Initialize inputs
initial begin
   clock = 1;
   bit_err = 0;
   start = 0;
   data_in = 0;
end

// Generate the clock
always #100 clock = ~clock;

// Simulate

// Look at the rising edge of the clock
always @(posedge clock) begin
   // On the first cycle, corrupt no bits
   // On each subsequent cycle, we corrupt each bit
   // from LSB to MSB
   // We then repeat the process
   if (bit_err)
      bit_err <= #`DEL bit_err << 1;
   else
      bit_err <= #`DEL 1;

   if (data_in == 0) begin
      if (start === 0)
         start <= #`DEL 1;
      else begin
         $display("\nSimulation complete - no errors\n");
         $finish;
      end
   end
   else begin
      if (data_out !== data_in) begin
         $display("\nERROR at time %0t:", $time);
         $display("    data_in = %h", data_in);
         $display("    data_out = %h\n", data_out);

         // Use $stop for debugging
         $stop;
      end
      if ((error !== 1) && (bit_err !== 0)) begin
         $display("\nERROR at time %0t:", $time);
         $display("error signal was not asserted\n");
```

```
        // Use $stop for debugging
        $stop;
     end
     else if ((error !== 0) && (bit_err === 0)) begin
        $display("\nERROR at time %0t:", $time);
        $display("error signal was asserted\n");

        // Use $stop for debugging
        $stop;
     end
  end

  // Create the next data input
  data_in <= #`DEL data_in+1;
end
endmodule    // ham_sim
```

The Checksum

A checksum is a simple but effective way of verifying a long sequence of data. All a checksum generator does is add each data word in the sequence on each clock cycle and output the inverted, cumulative sum. All of the carry bits simply overflow the adder and are discarded. This inverted sum is then stored at the end of the sequence of data. When the data is read back, the checksum verifier again adds each data word. When the checksum is passed through the verifier, the resulting output should be zero. If it is not zero, some data has been corrupted and the sequence must be re-read or discarded. Of course there is a small possibility that the data gets corrupted but ends up with the correct checksum. The probability of this happening, though, is very small.

20.1 IMPLEMENTATION CODE

The Behavioral Level code and the RTL code for the checksum generator and verifier is identical. To generate the checksum the internal accumulator must first be cleared. Each data word is then added to the accumulator that is then inverted to create the checksum that is stored at the end of the data sequence. To verify a checksum the accumulator must again be cleared and again each value is added. When the checksum is added, the output value should be zero.

```
/***********************************************************/
// MODULE:      Checksum generator and verifier
//
// FILE NAME:   checksum.v
// VERSION:     1.1
// DATE:        January 1, 2003
// AUTHOR:      Bob Zeidman, Zeidman Consulting
//
// CODE TYPE:   Behavioral and RTL
//
// DESCRIPTION: This module defines a checksum generator
// and verifier. To generate the checksum the internal
// accumulator must first be cleared. Each data word is then
// added to the accumulator which is then inverted to create
// the checksum. To verify a checksum the accumulator must
// again be cleared and again each value is added. When the
// checksum is added, the output value is compared to zero
// to see if the data sequence was correct.
//
/***********************************************************/

// DEFINES
`define DEL  1      // Clock-to-output delay. Zero
                    // time delays can be confusing
                    // and sometimes cause problems.
`define BITS 8      // Number of bits in the data word

// TOP MODULE
module Checksum(
     clk,
     reset,
     data,
     out,
     zero);

// PARAMETERS

// INPUTS
input              clk;      // Clock
input              reset;    // Synchronous reset
input [`BITS-1:0] data;      // Input data
```

```
// OUTPUTS
output [`BITS-1:0]    out;        // Checksum output
output              zero;       // Output is zero?

// INOUTS

// SIGNAL DECLARATIONS
wire                clk;
wire                reset;
wire [`BITS-1:0]    data;
wire [`BITS-1:0]    out;
wire                zero;
reg  [`BITS-1:0]    acc;         // Accumulator


// ASSIGN STATEMENTS
assign #`DEL zero = (acc == ~`BITS'h0) ? 1'b1 : 1'b0;
assign #`DEL out = ~acc;

// MAIN CODE

// Look at the rising edge of the clock
always @(posedge clk) begin
    if (reset) begin
        // Reset the accumulator
        acc <= #`DEL `BITS'h0;
    end
    else
        acc <= #`DEL acc + data;
end
endmodule    // Checksum
```

20.2 SIMULATION CODE

This simulation code creates a long sequence of random data words. The checksum module calculates the checksum. This simulation code then repeats the sequence, placing the checksum at the end, and checks that the resulting output of the checksum module is zero as expected. It then repeats the test, but corrupts one bit of one of the data words when repeating the sequence the second time. The simulation then checks that the resulting output of the checksum module is non-zero, signaling an error in the data. This sequence of tests is repeated for the number of tests specified at the beginning of the file.

```
/************************************************************/
// MODULE:        checksum simulation
```

```
//
// FILE NAME:      check_sim.v
// VERSION:        1.1
// DATE:           January 1, 2003
// AUTHOR:         Bob Zeidman, Zeidman Consulting
//
// CODE TYPE:      Simulation
//
// DESCRIPTION:    This module provides stimuli for simulating
// a checksum generator/verifier. First it generates a
// sequence of random data words and records the checksum.
// Then it repeats the sequence of data words with the
// checksum at the end. The checksum output should be zero.
// Then the simulation generates a new sequence of random
// data words and records the checksum. When it repeats
// this sequence with the checksum at the end, it corrupts
// one of the data words. The resulting output should be
// non-zero. It repeats the process for the number of tests
// that are specified.
//
/*********************************************************/

// DEFINES
`define BITS 8                // Number of bits in the data word
`define RBITS (`BITS+1)       // Used for generating random data
`define LENGTH 512            // Number of data words in the
                             // sequence
`define TESTS 16              // Number of tests to perform

// TOP MODULE
module check_sim();

// PARAMETERS

// INPUTS

// OUTPUTS

// INOUTS

// SIGNAL DECLARATIONS
reg                 clk;
reg                 reset;
reg    [`BITS-1:0]  data;
```

```
wire [`BITS-1:0]      out;
wire                  zero;

integer           cycle_count;   // Counts clock cycles
integer           test_count;    // Counts the test number

reg  [`BITS-1:0]     sum;             // Save the checksum

// ASSIGN STATEMENTS

// MAIN CODE

// Instantiate the checksum logic
Checksum checksum(
        .clk(clk),
        .reset(reset),
        .data(data),
        .out(out),
        .zero(zero));

// Initialize inputs
initial begin
    clk = 0;
    reset = 1;
    cycle_count = 0;
    test_count = 0;
end
// Generate the clock
always #100 clk = ~clk;

// Simulate
always @(negedge clk) begin
    // Check the zero signal
    if (((out !== `BITS'h0) && (zero !== 1'b0)) ||
        ((out === `BITS'h0) && (zero !== 1'b1))) begin
        $display("\nERROR at time %0t:", $time);
        $display("Zero flag is incorrect");
        $display("    checksum = %h", out);
        $display("    zero signal = %b\n", zero);

        // Use $stop for debugging
        $stop;
    end

    if (cycle_count !== 0) begin
        // Create random inputs
        data = {$random} % (`RBITS'h1 << `BITS);
```

```
     end

case (cycle_count)
    0:  begin

             // Test checksum
             if (out === ~`BITS'h0)
                 $display ("Reset is working");
             else begin
                 $display("\nERROR at time %0t:", $time);
                 $display("Reset is not working");
                 $display("    checksum = %h\n", out);

                 // Use $stop for debugging
                 $stop;
             end

             // Deassert the reset signal
             reset = 0;

             // Save the test_count, because the random
             // number generator changes it
             cycle_count = test_count;

             // Initialize random number generator
             data = $random(test_count);

             // Restore the test_count and cycle_count
             test_count = cycle_count;
             cycle_count = 0;

             // Create random inputs
             data = {$random} % (`RBITS'h1 << `BITS);
        end
    `LENGTH: begin
        // Save the checksum
        sum = out;
        // Assert reset
        reset = 1;
    end
    `LENGTH+1: begin
        // Test checksum
        if (out === ~`BITS'h0)
            $display ("Reset is working");
        else begin
            $display("\nERROR at time %0t:", $time);
```

```
            $display("Reset is not working");
            $display("    checksum = %h\n", out);

            // Use $stop for debugging
            $stop;
        end

        // Deassert the reset signal
        reset = 0;

        // Save the test_count, because the random
        // number generator changes it
        cycle_count = test_count;

        // Initialize random number generator
        data = $random(test_count);

        // Restore the test_count and cycle_count
        test_count = cycle_count;
        cycle_count = `LENGTH+1;

        // Create random inputs
        data = {$random} % (`RBITS'h1 << `BITS);
    end
`LENGTH+2: begin
    if (test_count & 1) begin
        $display("Corrupting data word");
        // Corrupt one bit of one data word
        data = data ^ `BITS'h1;
    end
end
2*`LENGTH+1: begin
    // Use the checksum as the last data word
    data = sum;
end
2*`LENGTH+2: begin
    // Test outputs
    if (test_count & 1) begin
        if ((out !== `BITS'h0) && (zero !== 1'b1))
            $display ("Checksum test #%d passed",
                test_count);
        else begin
            $display("\nERROR at time %0t:", $time);
            $display("Checksum is incorrect");
            $display("    zero flag    = %h", zero);
            $display("    checksum     = %h", out);
```

```
                    $display("     expected value = non-zero\n");
                    // Use $stop for debugging
                    $stop;
                end
        end
        else begin
            if ((out === `BITS'h0) && (zero === 1'b1))
                $display ("Checksum test #%d passed",
                    test_count);
            else begin
                $display("\nERROR at time %0t:", $time);
                $display("Checksum is incorrect");
                $display("     zero flag      = %h", zero);
                $display("     checksum       = %h", out);
                $display("     expected value = %h\n",
                    `BITS'h0);

                // Use $stop for debugging
                $stop;
            end
        end

        // Increment the test count
        test_count = test_count + 1;
        if (test_count === `TESTS) begin
            $display("\nSimulation complete - no errors\n");
            $finish;
        end

        // Initialize cycle_count so that it is
        // zero after incrementaion
        cycle_count = -1;

        // Assert the reset signal
        reset = 1;

    end
  endcase
  // Increment the cycle count
  cycle_count = cycle_count + 1;
end
endmodule     // check_sim
```

The Cyclic Redundancy Check (CRC)

The Cyclic Redundancy Check (CRC) is a method for checking a sequence of bits for errors. It is similar to the checksum, but more robust because there is a much smaller probability that the sequence can have errors that are not detected. The CRC is commonly used in data communications.

The mathematics of the CRC is beyond the scope of this textbook. From a practical point of view, it is important to know that the CRC is based on the Linear Feedback Shift Register (LFSR) which discussed in a previous chapter. The CRC uses the LFSR to take in the stream of bits of a given length, called a frame, and produce a Frame Check Sequence (FCS), which is simply a short sequence of bits. The FCS is then tacked onto the end of the frame when the data is transmitted. When the data is received, the entire frame plus FCS is put through an identical LFSR. The result should be all zeros. If not, an error has occurred.

To illustrate this, a five-bit LFSR is shown in Figure 21-1. The width of the LFSR must be the width of the FCS. This can be chosen arbitrarily. The longer the FCS, the greater the ability to detect errors. To create the FCS, shift the bit stream into the LFSR followed by zeros in the place of the FCS. The result in the register will be the FCS to append. When checking the FCS, shift the bit stream into the LFSR followed by the FCS. The result in the register should be all zeros. If not, an error has occurred.

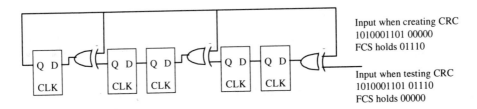

Input when creating CRC
1010001101 00000
FCS holds 01110

Input when testing CRC
1010001101 01110
FCS holds 00000

Figure 21-1 A cyclic redundancy check circuit (CRC).

21.1 BEHAVIORAL CODE

```
/*********************************************************/
// MODULE:        cyclic redundancy check (CRC)
//
// FILE NAME:     crc_beh.v
// VERSION:       1.1
// DATE:          January 1, 2003
// AUTHOR:        Bob Zeidman, Zeidman Consulting
//
// CODE TYPE:     Behavioral Level
//
// DESCRIPTION:   This module defines a cyclic redundancy
// check for error checking of sequences of bits. It is based
// on a linear feedback shift register (LFSR).
//
/*********************************************************/

// DEFINES
`define DEL   1      // Clock-to-output delay. Zero
                     // time delays can be confusing
                     // and sometimes cause problems.

                     // These are good tap values for 2 to 32 bits
`define TAP2  2'b11
`define TAP3  3'b101
`define TAP4  4'b1001
`define TAP5  5'b10010
`define TAP6  6'b100001
`define TAP7  7'b1000001
`define TAP8  8'b10001110
`define TAP9  9'b100001000
```

```
`define TAP10 10'b1000000100
`define TAP11 11'b10000000010
`define TAP12 12'b100000101001
`define TAP13 13'b1000000001101
`define TAP14 14'b10000000010101
`define TAP15 15'b100000000000001
`define TAP16 16'b1000000000010110
`define TAP17 17'b10000000000000100
`define TAP18 18'b100000000001000000
`define TAP19 19'b1000000000000010011
`define TAP20 20'b10000000000000000100
`define TAP21 21'b100000000000000000010
`define TAP22 22'b1000000000000000000001
`define TAP23 23'b10000000000000000010000
`define TAP24 24'b100000000000000000001101
`define TAP25 25'b1000000000000000000000100
`define TAP26 26'b10000000000000000000100011
`define TAP27 27'b100000000000000000000010011
`define TAP28 28'b1000000000000000000000000100
`define TAP29 29'b10000000000000000000000000010
`define TAP30 30'b100000000000000000000000101001
`define TAP31 31'b1000000000000000000000000000100
`define TAP32 32'b10000000000000000000000001100010

`define FCS 8        // Number of bits in the frame
`define TAPS `TAP8   // This must be the taps for the
                     // number of bits specified above

// TOP MODULE
module CRC(
       clk,
       reset,
       bit_in,
       fcs);

// PARAMETERS

// INPUTS
input                clk;     // Clock
input                reset;   // Synchronous reset
input                bit_in;  // Input bit stream

// OUTPUTS
output [`FCS-1:0]    fcs;     // FCS output

// INOUTS
```

```verilog
// SIGNAL DECLARATIONS
wire              clk;
wire              reset;
wire              bit_in;
reg [`FCS-1:0]    fcs;

// ASSIGN STATEMENTS

// MAIN CODE

// Look for the reset signal
always @(reset) begin
    // if reset gets deasserted, unforce the output
    if (~reset) begin
        deassign fcs;
        fcs = `FCS'h0;
    end
    // Wait for the rising edge of the clock
    @(posedge clk);
    // If reset is asserted at the clock, force the output
    // to the current input value (delayed by `DEL)
    if (reset) begin
        #`DEL assign fcs = `FCS'h0;
    end
end

// Look at the rising edge of clock for state transitions
always @(posedge clk) begin
    // XOR each tap bit with the most significant bit
    // and shift left. Also XOR the most significant
    // bit with the incoming bit
    if (fcs[`FCS-1]) begin
        fcs <= #`DEL (fcs ^ `TAPS) << 1;
        fcs[0] <= #`DEL ~bit_in;
    end
    else begin
        fcs <= #`DEL fcs << 1;
        fcs[0] <= #`DEL bit_in;
    end
end
endmodule    // CRC
```

21.2 RTL CODE

```
/***********************************************************/
// MODULE:        cyclic redundancy check (CRC)
//
// FILE NAME:     crc_rtl.v
// VERSION:       1.1
// DATE:          January 1, 2003
// AUTHOR:        Bob Zeidman, Zeidman Consulting
//
// CODE TYPE:     RTL
//
// DESCRIPTION:  This module defines a cyclic redundancy
// check for error checking of sequences of bits. It is based
// on a linear feedback shift register (LFSR).
//
/***********************************************************/

// DEFINES
`define DEL   1      // Clock-to-output delay. Zero
                     // time delays can be confusing
                     // and sometimes cause problems.

                  // These are good tap values for 2 to 32 bits
`define TAP2  2'b11
`define TAP3  3'b101
`define TAP4  4'b1001
`define TAP5  5'b10010
`define TAP6  6'b100001
`define TAP7  7'b1000001
`define TAP8  8'b10001110
`define TAP9  9'b100001000
`define TAP10 10'b1000000100
`define TAP11 11'b10000000010
`define TAP12 12'b100000101001
`define TAP13 13'b1000000001101
`define TAP14 14'b10000000010101
`define TAP15 15'b100000000000001
`define TAP16 16'b1000000000010110
`define TAP17 17'b10000000000000100
`define TAP18 18'b100000000001000000
`define TAP19 19'b1000000000000010011
`define TAP20 20'b10000000000000000100
`define TAP21 21'b100000000000000000010
`define TAP22 22'b1000000000000000000001
```

```
`define TAP23 23'b1000000000000000000010000
`define TAP24 24'b1000000000000000000001101
`define TAP25 25'b1000000000000000000000100
`define TAP26 26'b1000000000000000000100011
`define TAP27 27'b1000000000000000000010011
`define TAP28 28'b1000000000000000000000100
`define TAP29 29'b1000000000000000000000010
`define TAP30 30'b1000000000000000000000101001
`define TAP31 31'b1000000000000000000000000100
`define TAP32 32'b1000000000000000000000001100010

`define FCS 8        // Number of bits in the frame
`define TAPS `TAP8   // This must be the taps for the
                     // number of bits specified above

// TOP MODULE
module CRC(
      clk,
      reset,
      bit_in,
      fcs);

// PARAMETERS

// INPUTS
input               clk;       // Clock
input               reset;     // Synchronous reset
input               bit_in;    // Input bit stream

// OUTPUTS
output [`FCS-1:0]   fcs;       // FCS output

// INOUTS

// SIGNAL DECLARATIONS
wire                clk;
wire                reset;
wire                bit_in;
reg [`FCS-1:0]      fcs;

// ASSIGN STATEMENTS

// MAIN CODE

// Look at the rising edge of clock for state transitions
always @(posedge clk) begin
   if (reset) begin
      fcs <= #`DEL `FCS'h0;
   end
```

```
       else begin
          // XOR each tap bit with the most significant bit
          // and shift left. Also XOR the most significant
          // bit with the incoming bit
          if (fcs[`FCS-1]) begin
             fcs <= #`DEL (fcs ^ `TAPS) << 1;
             fcs[0] <= #`DEL ~bit_in;
          end
          else begin
             fcs <= #`DEL fcs << 1;
             fcs[0] <= #`DEL bit_in;
          end
       end
end
endmodule     // CRC
```

21.3 SIMULATION CODE

The simulation code creates a large, random, frame of data. The frame is then shifted through an LFSR to produce an FCS that is appended to the frame. In every even frame, a single bit is corrupted. Each frame, with the appended FCS, is then shifted through the LFSR once again. For uncorrupted frames, the LFSR is expected to contain zero at the end. For corrupted frames, the LFSR is expected to have a non-zero value.

```
/************************************************************/
// MODULE:        CRC simulation
//
// FILE NAME:     crc_sim.v
// VERSION:       1.1
// DATE:          January 1, 2003
// AUTHOR:        Bob Zeidman, Zeidman Consulting
//
// CODE TYPE:     Simulation
//
// DESCRIPTION:  This module provides stimuli for simulating
// a CRC generator/verifier. It creates a large, random,
// frame of data which is shifted through an LFSR to produce
// an FCS which is appended to the frame. In every even
// frame, a single bit is corrupted. Each frame, with the
// appended FCS, is then shifted through the LFSR again. For
// uncorrupted frames, the LFSR is expected to contain zero
// at the end. For corrupted frames, the LFSR is expected to
// have a non-zero value.
//
/************************************************************/
```

```verilog
// DEFINES
`define FCS 8                  // Number of bits in the fcs
`define FRAME 128              // Number of bytes in the frame
`define BFRAME `FRAME*8        // Number of bits in the frame
`define TOT_BITS `BFRAME+`FCS  // Total number of bits
                              // including frame and FCS
`define FRAME_CNT 16          // Number of frames to test

// TOP MODULE
module crc_sim();

// PARAMETERS

// INPUTS

// OUTPUTS

// INOUTS

// SIGNAL DECLARATIONS
reg                    clk;
reg                    reset;
wire                   bit_in;
wire [`FCS-1:0]        fcs;

integer                cycle_count; // Counts clock cycles
integer                frame_count; // Counts frames
reg  [`TOT_BITS-1:0]     frame_data;  // Frame data bits
reg                    gen_check;      // Generate/check CRC
                              // = 1 to generate CRC
                              // = 0 to check CRC
integer                i;          // Temporary variable

// ASSIGN STATEMENTS
assign bit_in = frame_data[cycle_count];

// MAIN CODE

// Instantiate the CRC logic
CRC crc(
     .clk(clk),
     .reset(reset),
     .bit_in(bit_in),
     .fcs(fcs));

// Initialize inputs
initial begin
   clk = 0;
```

```
    reset = 1;                      // Reset the FCS
    gen_check = 1;                  // Generate FCS
    cycle_count = `TOT_BITS - 1;
    frame_count = `FRAME_CNT;

    // Initialize random number generator
    $random(0);

    // Create random frame data of `FRAME bytes
    for (i = 0; i < `FRAME; i = i + 1) begin
        frame_data = (frame_data << 8) | ({$random} % 256);
    end
    // Then shift it left `FCS places
    frame_data = frame_data << `FCS;
end

// Generate the clock
always #100 clk = ~clk;

// Simulate
always @(negedge clk) begin
    // If reset is on, turn it off
    if (reset)
        reset = 0;
    else begin
        if (cycle_count === 0) begin
            if (gen_check) begin
                // Begin the CRC check
                gen_check = 0;
                cycle_count = `TOT_BITS - 1;

                // Put the FCS at the end of the data stream
                frame_data[`FCS-1:0] = fcs;

                // Corrupt one bit one every other test
                if ((frame_count & 1) === 0) begin
                    $display("Corrupting frame");

                    // Choose a random bit to corrupt
                    i = {$random} % (`TOT_BITS);
                    frame_data = frame_data ^ (`TOT_BITS'h1 << i);
                end

                // Reset the FCS
                reset = 1;
            end
            else begin
                if (((frame_count & 1) !== 0) &&
                    (fcs !== `FCS'h0)) begin
```

```
                    $display("\nERROR at time %0t:", $time);
                    $display("CRC produced %h instead of 0\n",
                        fcs, );

                    // Use $stop for debugging
                    $stop;
                end
                else if (((frame_count & 1) === 0) &&
                    (fcs === `FCS'h0)) begin
                    $display("\nERROR at time %0t:", $time);
                    $display("CRC passed a bad frame\n", fcs, );

                    // Use $stop for debugging
                    $stop;
                end
                else begin
                    $display("CRC check #%d passed",
                        `FRAME_CNT-frame_count);

                    // Reset the FCS
                    reset = 1;
                end

                if (frame_count === 0) begin
                 $display("\nSimulation complete - no errors\n");
                 $finish;
                end
                else begin
                    // Start the next frame
                    frame_count = frame_count - 1;
                    cycle_count = `TOT_BITS - 1;
                    gen_check = 1;

                    // Create random frame data of `FRAME bytes
                    for (i = 0; i < `FRAME; i = i + 1) begin
                        frame_data = (frame_data << 8) |
                            ({$random} % 256);
                    end
                    // Then shift it left `FCS places
                    frame_data = frame_data << `FCS;
                end
            end
        end

        // Decrement the cycle count
        cycle_count = cycle_count - 1;
    end
end
endmodule        // crc_sim
```

MEMORIES

*T*he following chapters describe various types of memories, from simple, standard types to special purpose memories. The Verilog code is given in these chapters to implement these memories in ASICs or FPGAs. Note that the actual gate level representation will vary greatly depending upon the architecture of the target chip. For example, many RAM-based FPGAs can implement memories using the on-chip RAM cells. Other FPGAs can only implement memories using registers, which require much more chip area and utilize more resources. Similarly, some ASICs include blocks of RAM which can be used to implement memories, while others do not.

In addition, your synthesis tool may be able to recognize memories and find the optimal gate level description, while other synthesis tools may automatically synthesis memories using registers, regardless of whether RAM is available. If you are designing memories and your chip architecture includes RAM, it is important to find a synthesis tool which can utilize the RAM to generate memories.

The Random Access Memory (RAM)

The Random Access Memory (RAM) is an important part of many designs. Now that complex systems are being designed on a chip, the accompanying system RAM is often moved onto the chip also. This chapter describes the Verilog code for the very simple RAM shown in Figure 22-1. Note that this RAM is asynchronous in that there is no common clock. Writing takes place with respect to the write signal, whereas reading takes place with respect to the output enable signal. This is about as simple an example of asynchronous RAM as you can have. For an example of a synchronous RAM, see the Dual Port RAM described in the next chapter.

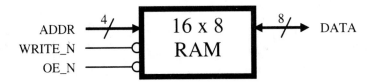

Figure 22-1 A simple 16 by 8 RAM.

22.1 IMPLEMENTATION CODE

Following is the Verilog code for a 16 deep RAM of 8-bit words. Due to the simplicity of the device, the behavioral and RTL versions of the code are identical.

```
/****************************************************************/
// MODULE:        RAM
//
// FILE NAME:     ram.v
// VERSION:       1.1
// DATE:          January 1, 2003
// AUTHOR:        Bob Zeidman, Zeidman Consulting
//
// CODE TYPE:     Behavioral and RTL
//
// DESCRIPTION:  This module defines a Random Access Memory.
//
/****************************************************************/

// DEFINES
`define DEL   1          // Clock-to-output delay. Zero
                         // time delays can be confusing
                         // and sometimes cause problems.

`define RAM_WIDTH 8      // Width of RAM (number of bits)
`define RAM_DEPTH 16     // Depth of RAM (number of bytes)
`define ADDR_SZ 4        // Number of bits required to
                         // represent the RAM address

// TOP MODULE
module Ram(
      data,
      address,
      write_n,
      oe_n);
```

```
// PARAMETERS

// INPUTS
input [`ADDR_SZ-1:0]      address;   // RAM address
input                     write_n;   // Write strobe (active low)
input                     oe_n;      // Output enable (active low)

// OUTPUTS

// INOUTS
inout [`RAM_WIDTH-1:0]  data;        // RAM data

// SIGNAL DECLARATIONS
wire [`ADDR_SZ-1:0]       address;
wire                      write_n;
wire                      oe_n;
wire [`RAM_WIDTH-1:0]   data;

                            // The RAM
reg  [`RAM_WIDTH-1:0]   mem [`RAM_DEPTH-1:0];

// ASSIGN STATEMENTS
assign #`DEL data = oe_n ? `RAM_WIDTH'bz : mem[address];

// MAIN CODE

// Look at the rising edge of the write signal
always @(posedge write_n) begin
   mem[address] = data;
end
endmodule      // Ram
```

22.2 SIMULATION CODE

This simulation code checks the RAM by writing unique values to each location. Once that has been done, it reads back each location and checks that the value is the same one that was previously written to it. Note that the data sequence is decrementing while the address sequence is incrementing. This is done in case the Verilog code for implementing the RAM has accidentally been written such that the address is used as the data. If this were to happen, then a test that wrote the address to each location would never uncover this bug.

```
/******************************************************/
// MODULE:         RAM simulation
//
// FILE NAME:      ram_sim.v
// VERSION:        1.1
```

```
// DATE:            January 1, 2003
// AUTHOR:          Bob Zeidman, Zeidman Consulting
//
// CODE TYPE:       Simulation
//
// DESCRIPTION: This module provides stimuli for simulating
// a Random Access Memory. It writes unique values to each
// location then reads each location back and checks for
// correctness.
//
/**********************************************************/

// DEFINES
`define DEL   1            // Clock-to-output delay. Zero
                           // time delays can be confusing
                           // and sometimes cause problems.

`define RAM_WIDTH 8        // Width of RAM (number of bits)
`define RAM_DEPTH 16       // Depth of RAM (number of bytes)
`define ADDR_SZ 4          // Number of bits required to
                           // represent the RAM address

// TOP MODULE
module ram_sim();

// PARAMETERS

// INPUTS

// OUTPUTS

// INOUTS

// SIGNAL DECLARATIONS
reg   [`ADDR_SZ-1:0]       address;
reg                        write_n;
reg                        oe_n;
wire [`RAM_WIDTH-1:0]      data;

reg  [`RAM_WIDTH-1:0]      data_in;     // Input data
reg  [`RAM_WIDTH-1:0]      data_exp;    // Expected output data

// ASSIGN STATEMENTS
assign #`DEL data = oe_n ? data_in : `RAM_WIDTH'bz;

// MAIN CODE

// Instantiate the counter
Ram ram(
      .data(data),
```

```
      .address(address),
      .write_n(write_n),
      .oe_n(oe_n));

// Initialize inputs
initial begin
   data_in = 0;
   address = 0;
   write_n = 1;
   oe_n = 1;

   // Start the action
   write_n <= #20 0;
end

// Simulate
// Write the RAM
always @(negedge write_n) begin
   // Bring write high to write to the RAM
   #10 write_n = 1;

   // Set up the address for the next write
   #10 address = address + 1;

   if (address === 0) begin
      // If the address is 0, we've written the entire RAM
      // Set up the reads
      oe_n <= #10 0;

      data_exp = 0;
   end
   else begin
      // Otherwise set up the data for the next write
      // We decrement data while incrementing address
      // so that we know we are writing the data, not
      // the address into memory
      data_in <= #10 data_in - 1;

      write_n <= #10 0;
   end
end

// Read the RAM
always @(negedge oe_n) begin
   // Read the data and compare
   #`DEL;
   #`DEL;
   if (data !== data_exp) begin
```

```verilog
        $display("\nERROR at time %0t:", $time);
        $display("   Data read     = %h", data);
        $display("   Data expected = %h\n", data_exp);

        // Use $stop for debugging
        $stop;
    end

    // Increment the address
    #10 address = address + 1;
    if (address === 0) begin
        // If the address is 0, we've read the entire RAM
        $display("\nSimulation complete - no errors\n");
        $finish;
    end

    // Decrement the expected data
    data_exp <= #10 data_exp - 1;

    // Set up the next rising edge of output enable
    oe_n <= #10 1;

    // Set up the next falling edge of output enable
    oe_n <= #30 0;
end
endmodule    // ram_sim
```

CHAPTER 23

The Dual Port RAM

The dual port RAM is a RAM that can be written and read simultaneously. This special type of RAM has two unidirectional data ports—an input port for writing data and an output port for reading data. Each port has its own data and address buses. The write port has a signal called WRITE to allow writing the data. The read port has a signal called READ to enable the data output. The particular dual port RAM examined in this chapter is synchronous and has a single clock for both ports, as shown in Figure 23-1. Both reading and writing data occur on the rising clock edge. For a description of an asynchronous RAM, see the previous chapter.

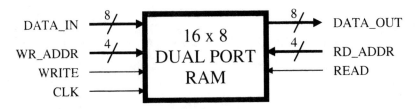

Figure 23-1 A simple 16 by 8 dual port RAM.

23.1 IMPLEMENTATION CODE

Following is the Verilog code for a 16 deep Dual Port RAM of 8-bit words. Due to the simplicity of the device, the behavioral and RTL versions of the code are identical.

```
/*******************************************************/
// MODULE:         Dual Port RAM
//
// FILE NAME:      dual.v
// VERSION:        1.1
// DATE:           January 1, 2003
// AUTHOR:         Bob Zeidman, Zeidman Consulting
//
// CODE TYPE:      Behavioral and RTL
//
// DESCRIPTION:  This module defines a Synchronous Dual Port
// Random Access Memory.
//
/*******************************************************/

// DEFINES
`define DEL   1          // Clock-to-output delay. Zero
                         // time delays can be confusing
                         // and sometimes cause problems.

`define RAM_WIDTH 8      // Width of RAM (number of bits)
`define RAM_DEPTH 16     // Depth of RAM (number of bytes)
`define ADDR_SZ 4        // Number of bits required to
                         // represent the RAM address

// TOP MODULE
module Dual(
      clk,
      data_in,
```

```
        rd_address,
        read,
        data_out,
        wr_address,
        write);

// PARAMETERS

// INPUTS
input                       clk;        // RAM clock
input [`RAM_WIDTH-1:0]  data_in;        // RAM data input
input [`ADDR_SZ-1:0]      rd_address;   // RAM read address
input                       read;       // Read control
input [`ADDR_SZ-1:0]      wr_address;   // RAM write address
input                       write;      // Write control

// OUTPUTS
output [`RAM_WIDTH-1:0] data_out;       // RAM data output

// INOUTS

// SIGNAL DECLARATIONS
wire                        clk;
wire [`RAM_WIDTH-1:0]   data_in;
wire [`ADDR_SZ-1:0]      rd_address;
wire                        read;
wire [`ADDR_SZ-1:0]      wr_address;
wire                        write;
reg  [`RAM_WIDTH-1:0]   data_out;

                                     // The RAM
reg [`RAM_WIDTH-1:0]        mem [`RAM_DEPTH-1:0];

// ASSIGN STATEMENTS

// MAIN CODE

// Look at the rising edge of the clock
always @(posedge clk) begin
   if (write)
      mem[wr_address] <= #`DEL data_in;

   if (read)
      data_out <= #`DEL mem[rd_address];
end
endmodule    // Dual
```

23.2 SIMULATION CODE

This simulation code writes unique values to each location and then reads each location back and checks for correctness. The reading and writing of the RAM overlap, so that the dual port nature of this RAM is tested.

```
/************************************************************/
// MODULE:         Dual Port RAM simulation
//
// FILE NAME:      dual_sim.v
// VERSION:        1.1
// DATE:           January 1, 2003
// AUTHOR:         Bob Zeidman, Zeidman Consulting
//
// CODE TYPE:      Simulation
//
// DESCRIPTION:  This module provides stimuli for simulating
// a Dual Port Random Access Memory. It writes unique values
// to each location then reads them back and checks for
// correctness. Read back of the RAM begins before writing
// has finished to check that both can occur simultaneously.
//
/************************************************************/

// DEFINES
`define DEL   1          // Clock-to-output delay. Zero
                         // time delays can be confusing
                         // and sometimes cause problems.

`define RAM_WIDTH 8      // Width of RAM (number of bits)
`define RAM_DEPTH 16     // Depth of RAM (number of bytes)
`define ADDR_SZ 4        // Number of bits required to
                         // represent the RAM address

// TOP MODULE
module dual_sim();

// PARAMETERS

// INPUTS

// OUTPUTS                                        .

// INOUTS

// SIGNAL DECLARATIONS
```

```
reg                         clk;
reg  [`ADDR_SZ-1:0]         rd_address;
reg  [`ADDR_SZ-1:0]         wr_address;
reg                         read;
reg                         write;
reg  [`RAM_WIDTH-1:0]       data_in;
wire [`RAM_WIDTH-1:0]       data_out;

reg  [`RAM_WIDTH-1:0]       data_exp;     // Expected output data
integer                     cyc_count;    // Cycle counter

// ASSIGN STATEMENTS

// MAIN CODE

// Instantiate the counter
Dual dual(
        .clk(clk),
        .data_in(data_in),
        .rd_address(rd_address),
        .read(read),
        .data_out(data_out),
        .wr_address(wr_address),
        .write(write));

// Initialize inputs
initial begin
    data_in = 0;
    data_exp = 0;
    rd_address = 0;
    wr_address = 0;
    clk = 1;
    cyc_count = 0;
    write = 1;       // Start writing
    read = 0;        // Start reading later
end

// Generate the clock
always #100 clk = ~clk;

// Simulate
// Write the RAM
always @(posedge clk) begin
    // Give a delay for outputs to settle
    #`DEL;
    #`DEL;
```

```verilog
    if (write) begin
        // Set up the write address for the next write
        wr_address = wr_address + 1;
        if (wr_address === 0) begin
            // If the address is 0, we've written the entire RAM
            // Deassert the write control
            write = 0;
        end
        else begin
            // Otherwise set up the data for the next write
            // We decrement data while incrementing address
            // so that we know we are writing the data, not
            // the address into memory
            data_in = data_in - 1;
        end
    end

    if (read) begin
        // Read the data and compare
        if (data_out !== data_exp) begin
            $display("\nERROR at time %0t:", $time);
            $display("    Data read     = %h", data_out);
            $display("    Data expected = %h\n", data_exp);

            // Use $stop for debugging
            $stop;
        end

        // Increment the read address
        rd_address = rd_address + 1;
        if (rd_address === 0) begin
            // If the address is 0, we've read the entire RAM
            $display("\nSimulation complete - no errors\n");
            $finish;
        end

        // Decrement the expected data
        data_exp <= data_exp - 1;
    end

    // Increment the cycle counter
    cyc_count = cyc_count + 1;

    // Start reading at some point
    if (cyc_count == (`RAM_DEPTH/2 + 2)) begin
        // Assert the read control
        read = 1;
    end
end
endmodule     // dual_sim
```

The Synchronous FIFO

The FIFO (First In First Out) is a type of memory that is commonly used to buffer data that is being transferred between different systems or different parts of a system, which are operating at different speeds or with different delays. The FIFO allows the transmitter to send data while the receiver is not ready. The data then fills up the FIFO memory until the receiver begins unloading it. An overflow occurs when the transmitter fills up the FIFO and attempts to store more data before the receiver has read the data out. An underflow occurs when the receiver attempts to read data from the FIFO, but the transmitter has not yet placed any data into it. Full and empty signals are used by the logic to throttle the transmitter and receiver, respectively, in order to avoid these conditions. A half-full signal and almost full and almost empty signals are also used to throttle the devices in cases where data is transferred very quickly and the throttling logic needs more time to inform the transmitter or receiver to stop.

Figure 24-1 shows the functionality of the FIFO. The transmitter puts data into the FIFO, like filling a bucket with water. The newest data is on top, whereas the oldest data is on the bottom. The receiver gets data out of the FIFO like emptying the bucket using a faucet.

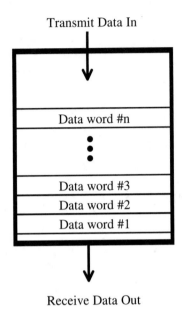

Figure 24-1 FIFO functional diagram.

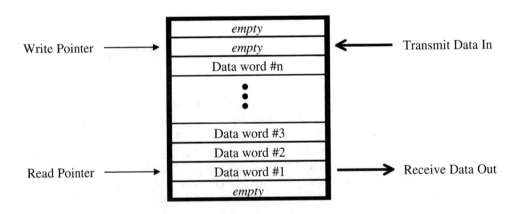

Figure 24-2 FIFO implementation.

Figure 24-2 shows our actual hardware implementation of the FIFO. It is implemented as a ring buffer where two pointers are used to keep track of the top and bottom of the FIFO. The write pointer points to the next available location to write data—the top of the FIFO. The read pointer points to the next data to be read—the bottom of the FIFO. When a pointer is incremented past the end of the buffer, it is set back to the beginning of the buffer.

In this chapter, we specifically examine the synchronous FIFO. By this, I mean that both the read and write sides of the FIFO are synchronized to the same clock. This greatly simplifies the design. However, many designs require a synchronizing FIFO that is described in the next chapter. For a synchronizing FIFO, the input and output sides are synchronized to different clocks. It is the job of the FIFO to coordinate the reads and writes and essentially synchronize the receive logic to the transmit logic.

A counter is also used to keep track of the number of data words currently in the FIFO. This is necessary, because when the write pointer and the read pointer are pointing to the same location, we cannot otherwise tell if the FIFO is completely empty or completely full. The counter can also be decoded to give the full, almost full, half full, almost empty, and empty signals, if needed.

24.1 BEHAVIORAL CODE

Following is the behavioral code for a 15 deep FIFO of 8-bit words. Notice that there is a *$display* statement in the code. This statement gets executed whenever the FIFO overflows or underflows. This demonstrates how the hardware code can check itself for illegal conditions and report them.

```
/*****************************************************/
// MODULE:         Synchronous FIFO
//
// FILE NAME:      sfifo_beh.v
// VERSION:        1.1
// DATE:           January 1, 2003
// AUTHOR:         Bob Zeidman, Zeidman Consulting
//
// CODE TYPE:      Behavioral Level
//
// DESCRIPTION:  This module defines a Synchronous FIFO. The
// FIFO memory is implemented as a ring buffer. The read
// pointer points to the beginning of the buffer, while the
// write pointer points to the end of the buffer.
//
/*****************************************************/

// DEFINES
`define DEL   1          // Clock-to-output delay. Zero
                         // time delays can be confusing
                         // and sometimes cause problems.

`define FIFO_DEPTH 15    // Depth of FIFO (number of bytes)
`define FIFO_HALF 8      // Half depth of FIFO
                         // (this avoids rounding errors)
```

```verilog
`define FIFO_BITS 4        // Number of bits required to
                          // represent the FIFO size
`define FIFO_WIDTH 8       // Width of FIFO data

// TOP MODULE
module Sfifo(
        clock,
        reset_n,
        data_in,
        read_n,
        write_n,
        data_out,
        full,
        empty,
        half);

// PARAMETERS
// INPUTS
input                     clock;    // Clock input
input                     reset_n;  // Active low reset
input [`FIFO_WIDTH-1:0]   data_in;  // Data input to FIFO
input                     read_n;   // Read FIFO (active low)
input                     write_n;  // Write FIFO (active low)

// OUTPUTS
output [`FIFO_WIDTH-1:0]  data_out; // FIFO output data
output                    full;     // FIFO is full
output                    empty;    // FIFO is empty
output                    half;     // FIFO is half full
                                    // or more

// INOUTS

// SIGNAL DECLARATIONS
wire                      clock;
wire                      reset_n;
wire [`FIFO_WIDTH-1:0]    data_in;
wire                      read_n;
wire                      write_n;
reg  [`FIFO_WIDTH-1:0]    data_out;
wire                      full;
wire                      empty;
wire                      half;

                          // The FIFO memory.
reg [`FIFO_WIDTH-1:0]     fifo_mem[0:`FIFO_DEPTH-1];
                          // How many locations in the FIFO
                          // are occupied?
```

```
reg [`FIFO_BITS-1:0]              counter;
                        // FIFO read pointer points to
                        // the location in the FIFO to
                        // read from next
reg [`FIFO_BITS-1:0]            rd_pointer;
                        // FIFO write pointer points to
                        // the location in the FIFO to
                        // write to next
reg [`FIFO_BITS-1:0]            wr_pointer;


// ASSIGN STATEMENTS
assign #`DEL full = (counter == `FIFO_DEPTH) ? 1'b1 : 1'b0;
assign #`DEL empty = (counter == 0) ? 1'b1 : 1'b0;
assign #`DEL half = (counter >= `FIFO_HALF) ? 1'b1 : 1'b0;

// MAIN CODE

// Look at the edges of reset_n
always @(reset_n) begin
    if (~reset_n) begin
        // Reset the FIFO pointer
        #`DEL;
        assign rd_pointer = `FIFO_BITS'b0;
        assign wr_pointer = `FIFO_BITS'b0;
        assign counter = `FIFO_BITS'b0;
    end
    else begin
        #`DEL;
        deassign rd_pointer;
        deassign wr_pointer;
        deassign counter;
    end
end

// Look at the rising edge of the clock
always @(posedge clock) begin
    if (~read_n) begin
        // Check for FIFO underflow
        if (counter == 0) begin
            $display("\nERROR at time %0t:", $time);
            $display("FIFO Underflow\n");

            // Use $stop for debugging
            $stop;
        end// If we are doing a simultaneous read and write,
```

```verilog
            // there is no change to the counter
            if (write_n) begin
               // Decrement the FIFO counter
               counter <= #`DEL counter - 1;
            end
            // Output the data
            data_out <= #`DEL fifo_mem[rd_pointer];

            // Increment the read pointer
            // Check if the read pointer has gone beyond the
            // depth of the FIFO. If so, set it back to the
            // beginning of the FIFO
            if (rd_pointer == `FIFO_DEPTH-1)
               rd_pointer <= #`DEL `FIFO_BITS'b0;
            else
               rd_pointer <= #`DEL rd_pointer + 1;
         end
         if (~write_n) begin
            // Check for FIFO overflow
            if (counter >= `FIFO_DEPTH) begin
               $display("\nERROR at time %0t:", $time);
               $display("FIFO Overflow\n");

               // Use $stop for debugging
               $stop;
            end

            // If we are doing a simultaneous read and write,
            // there is no change to the counter
            if (read_n) begin
               // Increment the FIFO counter
               counter <= #`DEL counter + 1;
            end

            // Store the data
            fifo_mem[wr_pointer] <= #`DEL data_in;

            // Increment the write pointer
            // Check if the write pointer has gone beyond the
            // depth of the FIFO. If so, set it back to the
            // beginning of the FIFO
            if (wr_pointer == `FIFO_DEPTH-1)
               wr_pointer <= #`DEL `FIFO_BITS'b0;
            else
               wr_pointer <= #`DEL wr_pointer + 1;
         end
      end
endmodule     // Sfifo
```

24.2 RTL CODE

Following is the RTL code for a 15 deep FIFO of 8-bit words. Notice that there is a *$display* statement in this code also. This statement gets executed whenever the FIFO overflows or underflows. This demonstrates how the hardware code can check itself for illegal conditions and report them. The synthesis tools knows that a *$display* statement does not represent real hardware. It will be ignored during synthesis and will not affect the resulting gate level implementation.

Also note that all devices that are not affected by reset must have their own always block. Otherwise the synthesis software will add unnecessary reset logic to the hardware, producing a size and speed penalty. The FIFO memory is one such device that is not affected by reset, since we are not interested in its initial values.

```
/**********************************************************/
// MODULE:        Synchronous FIFO
//
// FILE NAME:     sfifo_rtl.v
// VERSION:       1.1
// DATE:          January 1, 2003
// AUTHOR:        Bob Zeidman, Zeidman Consulting
//
// CODE TYPE:     Register Transfer Level
//
// DESCRIPTION:  This module defines a Synchronous FIFO. The
// FIFO memory is implemented as a ring buffer. The read
// pointer points to the beginning of the buffer, while the
// write pointer points to the end of the buffer. Note that
// in this RTL version, the memory has one more location than
// the FIFO needs in order to calculate the FIFO count
// correctly.
//
/**********************************************************/

// DEFINES
`define DEL   1          // Clock-to-output delay. Zero
                         // time delays can be confusing
                         // and sometimes cause problems.

`define FIFO_DEPTH 15    // Depth of FIFO (number of bytes)
`define FIFO_HALF 8      // Half depth of FIFO
                         // (this avoids rounding errors)
`define FIFO_BITS 4      // Number of bits required to
                         // represent the FIFO size
`define FIFO_WIDTH 8     // Width of FIFO data
```

```
// TOP MODULE
module Sfifo(
       clock,
       reset_n,
       data_in,
       read_n,
       write_n,
       data_out,
       full,
       empty,
       half);

// PARAMETERS

// INPUTS
input                           clock;     // Clock input
input                           reset_n;   // Active low reset
input [`FIFO_WIDTH-1:0]         data_in;   // Data input to FIFO
input                           read_n;    // Read FIFO (active low)
input                           write_n;   // Write FIFO (active low)

// OUTPUTS
output [`FIFO_WIDTH-1:0]        data_out;  // FIFO output data
output                          full;      // FIFO is full
output                          empty;     // FIFO is empty
output                          half;      // FIFO is half full
                                           // or more

// INOUTS

// SIGNAL DECLARATIONS
wire                            clock;
wire                            reset_n;
wire  [`FIFO_WIDTH-1:0]         data_in;
wire                            read_n;
wire                            write_n;
reg   [`FIFO_WIDTH-1:0]         data_out;
wire                            full;
wire                            empty;
wire                            half;

                      // The FIFO memory.
reg [`FIFO_WIDTH-1:0]     fifo_mem[0:`FIFO_DEPTH-1];
                      // How many locations in the FIFO
                      // are occupied?
reg [`FIFO_BITS-1:0]          counter;
                      // FIFO read pointer points to
                      // the location in the FIFO to
```

```
                            // read from next
reg ['FIFO_BITS-1:0]            rd_pointer;

                            // FIFO write pointer points to
                            // the location in the FIFO to
                            // write to next
reg ['FIFO_BITS-1:0]            wr_pointer;

// ASSIGN STATEMENTS
assign #`DEL full = (counter == `FIFO_DEPTH) ? 1'b1 : 1'b0;
assign #`DEL empty = (counter == 0) ? 1'b1 : 1'b0;
assign #`DEL half = (counter >= `FIFO_HALF) ? 1'b1 : 1'b0;

// MAIN CODE

// This block contains all devices affected by the clock
// and reset inputs
always @(posedge clock or negedge reset_n ) begin
    if (~reset_n) begin
        // Reset the FIFO pointer
        rd_pointer <= #`DEL `FIFO_BITS'b0;
        wr_pointer <= #`DEL `FIFO_BITS'b0;
        counter <= #`DEL `FIFO_BITS'b0;
    end
    else begin
        if (~read_n) begin
            // Check for FIFO underflow
            if (counter == 0) begin
                $display("\nERROR at time %0t:", $time);
                $display("FIFO Underflow\n");

                // Use $stop for debugging
                $stop;
            end

            // If we are doing a simultaneous read and write,
            // there is no change to the counter
            if (write_n) begin
                // Decrement the FIFO counter
                counter <= #`DEL counter - 1;
            end

            // Increment the read pointer
            // Check if the read pointer has gone beyond the
            // depth of the FIFO. If so, set it back to the
            // beginning of the FIFO
            if (rd_pointer == `FIFO_DEPTH-1)
                rd_pointer <= #`DEL `FIFO_BITS'b0;
```

```verilog
            else
                rd_pointer <= #`DEL rd_pointer + 1;
        end
        if (~write_n) begin
            // Check for FIFO overflow
            if (counter >= `FIFO_DEPTH) begin
                $display("\nERROR at time %0t:", $time);
                $display("FIFO Overflow\n");

                // Use $stop for debugging
                $stop;
            end

            // If we are doing a simultaneous read and write,
            // there is no change to the counter
            if (read_n) begin
                // Increment the FIFO counter
                counter <= #`DEL counter + 1;
            end

            // Increment the write pointer
            // Check if the write pointer has gone beyond the
            // depth of the FIFO. If so, set it back to the
            // beginning of the FIFO
            if (wr_pointer == `FIFO_DEPTH-1)
                wr_pointer <= #`DEL `FIFO_BITS'b0;
            else
                wr_pointer <= #`DEL wr_pointer + 1;
        end
    end
end

// This block contains all devices affected by the clock
// but not reset
always @(posedge clock) begin
    if (~read_n) begin
        // Output the data
        data_out <= #`DEL fifo_mem[rd_pointer];
    end
    if (~write_n) begin
        // Store the data
        fifo_mem[wr_pointer] <= #`DEL data_in;
    end
end
endmodule     // Sfifo
```

24.3 SIMULATION CODE

The simulation code for the FIFO is shown below. Note that the simulation uses the defined `fifo_depth that is also used in the code for the hardware function. In a real design, it is useful to keep all of the *defines* in a separate file that can be included in each Verilog code file. In that way, when the value of the *define* is changed, for any reason, that change is done once in the common file, and it affects each file where it is used in the code.

The simulation begins by writing quickly to the FIFO while reading slowly. This fills up the FIFO. Once the FIFO is filled, it changes the frequency of the reads and writes. Writing slowly and reading quickly, the FIFO empties and the simulation ends.

```
/*********************************************************/
// MODULE:      Synchronous FIFO simulation
//
// FILE NAME:   sfifo_sim.v
// VERSION:     1.1
// DATE:        January 1, 2003
// AUTHOR:      Bob Zeidman, Zeidman Consulting
//
// CODE TYPE:   Simulation
//
// DESCRIPTION: This module provides stimuli for simulating
// a Synchronous FIFO. It begins by writing quickly to the
// FIFO while reading slowly. This fills up the FIFO. Once
// the FIFO is filled, it changes the frequency of the reads
// and writes. Writing slowly and reading quickly, the FIFO
// empties and the simulation ends.
//
/*********************************************************/

// DEFINES
`define DEL   1        // Clock-to-output delay. Zero
                       // time delays can be confusing
                       // and sometimes cause problems.

`define FIFO_DEPTH 15  // Depth of FIFO (number of bytes)
`define FIFO_HALF 8    // Half depth of FIFO
                       // (this avoids rounding errors)
`define FIFO_WIDTH 8     // Width of FIFO data

// TOP MODULE
module sfifo_sim();

// PARAMETERS
```

```
// INPUTS

// OUTPUTS

// INOUTS

// SIGNAL DECLARATIONS
reg                        clock;
reg                        clr_n;
reg   [`FIFO_WIDTH-1:0]    in_data;
reg                        read_n;
reg                        write_n;
wire  [`FIFO_WIDTH-1:0]    out_data;
wire                       full;
wire                       empty;
wire                       half;

integer                    fifo_count;   // Keep track of the number
                                         // of bytes in the FIFO
reg [`FIFO_WIDTH-1:0]      exp_data;     // The expected data
                                         // from the FIFO
reg                        fast_read;      // Read at high frequency
reg                        fast_write;   // Write at high frequency
reg                        filled_flag;  // The FIFO has filled
                                         // at least once
reg                        cycle_count;  // Count the cycles

// ASSIGN STATEMENTS

// MAIN CODE

// Instantiate the counter
Sfifo sfifo(
        .clock(clock),
        .reset_n(clr_n),
        .data_in(in_data),
        .read_n(read_n),
        .write_n(write_n),
        .data_out(out_data),
        .full(full),
        .empty(empty),
        .half(half));

// Initialize inputs
initial begin
    in_data = 0;
    exp_data = 0;
```

```
        fifo_count = 0;
        read_n = 1;
        write_n = 1;
        filled_flag = 0;
        cycle_count = 0;
        clock = 1;

        // Write quickly to the FIFO
        fast_write = 1;
        // Read slowly from the FIFO
        fast_read = 0;

        // Reset the FIFO
        clr_n = 1;
        #20 clr_n = 0;
        #20 clr_n = 1;

        // Check that the status outputs are correct
        if (empty !== 1) begin
            $display("\nERROR at time %0t:", $time);
            $display("After reset, empty status not asserted\n");

            // Use $stop for debugging
            $stop;
        end
        if (full !== 0) begin
            $display("\nERROR at time %0t:", $time);
            $display("After reset, full status is asserted\n");

            // Use $stop for debugging
            $stop;
        end
        if (half !== 0) begin
            $display("\nERROR at time %0t:", $time);
            $display("After reset, half status is asserted\n");

            // Use $stop for debugging
            $stop;
        end
end

// Generate the clock
always #100 clock = ~clock;

// Simulate
always @(posedge clock) begin
    // Adjust the count if there is a write but no read
    // or a read but no write
```

```verilog
    if (~write_n && read_n)
        fifo_count = fifo_count + 1;
    else if (~read_n && write_n)
        fifo_count = fifo_count - 1;
end

always @(negedge clock) begin
    // Check the read data
    if (~read_n && (out_data !== exp_data)) begin
        $display("\nERROR at time %0t:", $time);
        $display("    Expected data out = %h", exp_data);
        $display("    Actual data out   = %h\n", out_data);

        // Use $stop for debugging
        $stop;
    end

    // Check whether to assert write_n
    // Do not write the FIFO if it is full
    if ((fast_write || (cycle_count & 1'b1)) &&
            ~full) begin
        write_n = 0;

        // Set up the data for the next write
        in_data = in_data + 1;
    end
    else
        write_n = 1;

    // Check whether to assert read_n
    // Do not read the FIFO if it is empty
    if ((fast_read || (cycle_count & 1'b1)) &&
            ~empty) begin
        read_n = 0;

        // Increment the expected data
        exp_data = exp_data + 1;
    end
    else
        read_n = 1;

    // When the FIFO is full, begin reading faster
    // than writing to empty it
    if (full) begin
        fast_read = 1;
        fast_write = 0;
```

```
        // Set the flag that the FIFO has been filled
        filled_flag = 1;
    end

    // When the FIFO has been filled then emptied,
    // we are done
    if (filled_flag && empty) begin
        $display("\nSimulation complete - no errors\n");
        $finish;
    end

    // Increment the cycle count
    cycle_count = cycle_count + 1;
end

// Check all of the status signals with each change
// of fifo_count
always @(fifo_count) begin
    // Wait a moment to evaluate everything
    #`DEL;
    #`DEL
    #`DEL;

    case (fifo_count)
        0: begin
            if ((empty !== 1) || (half !== 0) ||
                    (full !== 0)) begin
                $display("\nERROR at time %0t:", $time);
                $display("    fifo_count = %h", fifo_count);
                $display("    empty = %b", empty);
                $display("    half  = %b", half);
                $display("    full  = %b\n", full);

                // Use $stop for debugging
                $stop;
            end

            if (filled_flag === 1) begin
                // The FIFO has filled and emptied
                $display("\nSimulation complete - no errors\n");
                $finish;
            end
        end
        `FIFO_HALF: begin
            if ((empty !== 0) || (half !== 1) ||
                    (full !== 0)) begin
                $display("\nERROR at time %0t:", $time);
```

```
            $display("    fifo_count = %h", fifo_count);
            $display("    empty = %b", empty);
            $display("    half  = %b", half);
            $display("    full  = %b\n", full);

            // Use $stop for debugging
            $stop;
        end
    end
    `FIFO_DEPTH: begin
        if ((empty !== 0) || (half !== 1) ||
             (full !== 1)) begin
            $display("\nERROR at time %0t:", $time);
            $display("    fifo_count = %h", fifo_count);
            $display("    empty = %b", empty);
            $display("    half  = %b", half);
            $display("    full  = %b\n", full);

            // Use $stop for debugging
            $stop;
        end

        // The FIFO has filled, so set the flag
        filled_flag = 1;

        // Once the FIFO has filled, empty it
        // Write slowly to the FIFO
        fast_write = 0;
        // Read quickly from the FIFO
        fast_read = 1;
    end
    default: begin
        if ((empty !== 0) || (full !== 0)) begin
            $display("\nERROR at time %0t:", $time);
            $display("    fifo_count = %h", fifo_count);
            $display("    empty = %b", empty);
            $display("    half  = %b", half);
            $display("    full  = %b\n", full);

            // Use $stop for debugging
            $stop;
        end
        if (((fifo_count < `FIFO_HALF) &&
              (half === 1)) ||
             ((fifo_count >= `FIFO_HALF) &&
              (half === 0))) begin
            $display("\nERROR at time %0t:", $time);
```

```
            $display("    fifo_count = %h", fifo_count);
            $display("    empty = %b", empty);
            $display("    half  = %b", half);
            $display("    full  = %b\n", full);

            // Use $stop for debugging
            $stop;
        end
      end
   endcase
end

endmodule    // sfifo_sim
```

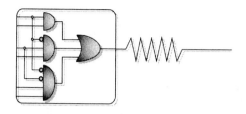

The Synchronizing FIFO

The Synchronizing FIFO is a memory buffer similar to the Synchronous FIFO described in the previous chapter. Unlike a Synchronous FIFO that has a single clock, the Synchronizing FIFO uses two clocks, one for the receiver logic and one for the transmitter logic. These clocks are asynchronous with respect to each other. That means that there is no defined phase relationship between them. The job of the Synchronizing FIFO is to synchronize the two sets of logic without losing or corrupting data. For this reason, this type of FIFO is much more complex to design. Note that this design has status bits (full, empty, half) that are generated using combinatorial logic of signals clocked off both clocks. These signals are therefore asynchronous with respect to both clocks. Before using these status bits, they will need to be synchronized to one of the two clocks in order to reduce the chances of metastability.

There are different ways of implementing FIFO interfaces, but I chose to just simply have a read signal and a write signal that are asynchronous with respect to each other. These signals are the two clocks. The data is read or written on the falling edge of each signal.

25.1 BEHAVIORAL CODE

Following is the behavioral code for a 15 deep FIFO of 8-bit words. As with the Synchronous FIFO described in the previous chapter, a counter is used to keep track of the number of data words currently in the FIFO. This is necessary, because when the write pointer and the read pointer are pointing to the same location, we cannot otherwise tell if the FIFO is completely empty of completely full. The counter is also decoded to give the full, almost full, half full, almost empty, and empty signals, if needed.

```
/****************************************************/
// MODULE:          Synchronizing FIFO
//
// FILE NAME:       afifo_beh.v
// VERSION:         1.1
// DATE:            January 1, 2003
// AUTHOR:          Bob Zeidman, Zeidman Consulting
//
// CODE TYPE:       Behavioral Level
//
// DESCRIPTION:  This module defines a Synchronizing FIFO.
// The FIFO memory is implemented as a ring buffer. The read
// pointer points to the beginning of the buffer, while the
// write pointer points to the end of the buffer.
//
/****************************************************/

// DEFINES
`define DEL   1          // Clock-to-output delay. Zero
                         // time delays can be confusing
                         // and sometimes cause problems.

`define FIFO_DEPTH 15    // Depth of FIFO (number of bytes)
`define FIFO_HALF 8      // Half depth of FIFO
                         // (this avoids rounding errors)
`define FIFO_BITS 4      // Number of bits required to
                         // represent the FIFO size
`define FIFO_WIDTH 8// Width of FIFO data

// TOP MODULE
module Afifo(
      reset_n,
      data_in,
      read_n,
      write_n,
      data_out,
      full,
```

```
              empty,
              half);

// PARAMETERS

// INPUTS
input                         reset_n;   // Active low reset
input [`FIFO_WIDTH-1:0]       data_in;   // Data input to FIFO
input                         read_n;    // Read FIFO (active low)
input                         write_n;   // Write FIFO (active low)

// OUTPUTS
output [`FIFO_WIDTH-1:0]      data_out;  // FIFO data output
output                        full;      // FIFO is full
output                        empty;     // FIFO is empty
output                        half;      // FIFO is half full
                                         // or more

// INOUTS

// SIGNAL DECLARATIONS
wire                          reset_n;
wire [`FIFO_WIDTH-1:0]        data_in;
wire                          read_n;
wire                          write_n;
reg  [`FIFO_WIDTH-1:0]        data_out;
wire                          full;
wire                          empty;
wire                          half;

                         // The FIFO memory.
reg [`FIFO_WIDTH-1:0]        fifo_mem[0:`FIFO_DEPTH-1];
                         // How many locations in the FIFO
                         // are occupied?
reg [`FIFO_BITS-1:0]          counter;
                         // FIFO read pointer points to
                         // the location in the FIFO to
                         // read from next
reg [`FIFO_BITS-1:0]          rd_pointer;
                         // FIFO write pointer points to
                         // the location in the FIFO to
                         // write to next
reg [`FIFO_BITS-1:0]          wr_pointer;

// ASSIGN STATEMENTS
assign #`DEL full = (counter == `FIFO_DEPTH) ? 1'b1 : 1'b0;
assign #`DEL empty = (counter == 0) ? 1'b1 : 1'b0;
```

```verilog
assign #`DEL half = (counter >= `FIFO_HALF) ? 1'b1 : 1'b0;

// MAIN CODE

// Look at the edges of reset_n
always @(reset_n) begin
    if (~reset_n) begin
        // Reset the FIFO pointer
        #`DEL;
        assign rd_pointer = `FIFO_BITS'b0;
        assign wr_pointer = `FIFO_BITS'b0;
        assign counter = `FIFO_BITS'b0;
    end
    else begin
        #`DEL;
        deassign rd_pointer;
        deassign wr_pointer;
        deassign counter;
    end
end

// Look at the falling edge of the read signal
always @(negedge read_n) begin
    // Check for FIFO underflow
    if (counter == 0) begin
        $display("\nERROR at time %0t:", $time);
        $display("FIFO Underflow\n");

        // Use $stop for debugging
        $stop;
    end

    // Decrement the FIFO counter
    counter = counter - 1;

    // Output the data
    data_out <= #`DEL fifo_mem[rd_pointer];

    // Increment the read pointer
    // Check if the read pointer has gone beyond the
    // depth of the FIFO. If so, set it back to the
    // beginning of the FIFO
    if (rd_pointer == `FIFO_DEPTH-1)
        rd_pointer <= #`DEL `FIFO_BITS'b0;
    else
        rd_pointer <= #`DEL rd_pointer + 1;
end
```

```
// Look at the falling edge of the write signal
always @(negedge write_n) begin
    // Check for FIFO overflow
    if (counter >= `FIFO_DEPTH) begin
        $display("\nERROR at time %0t:", $time);
        $display("FIFO Overflow\n");

        // Use $stop for debugging
        $stop;
    end

    // Increment the FIFO counter
    counter = counter + 1;

    // Store the data
    fifo_mem[wr_pointer] <= #`DEL data_in;

    // Increment the write pointer
    // Check if the write pointer has gone beyond the
    // depth of the FIFO. If so, set it back to the
    // beginning of the FIFO
    if (wr_pointer == `FIFO_DEPTH-1)
        wr_pointer <= #`DEL `FIFO_BITS'b0;
    else
        wr_pointer <= #`DEL wr_pointer + 1;
end
endmodule      // Afifo
```

25.2 RTL CODE

Following is the RTL code for a 15 deep FIFO of 8-bit words. This code is somewhat trickier than before. In order to synthesize correctly, the FIFO counter cannot have its value change in two different *always* blocks. This would represent a flip-flop with two different clock inputs—something that does not exist. If this device did exist, it would have metastability problems operating with two asynchronous clocks. Instead, the counter is simply derived from the read and write pointers, using combinatorial logic. This presents a new problem. When the read pointer is equal to the write pointer, do we have an empty FIFO or a full one? There is no way to know. The solution is to use a memory that has one more location than necessary. In our example of a 15 location FIFO, we use a 16 location memory. We consider the FIFO full when 15 out of the 16 locations are occupied with unread data. This wastes some memory, but is the only practical solution to the problem.

Also note that all devices that are not affected by reset must have their own always block. Otherwise the synthesis software will add unnecessary reset logic to the hardware, producing a size and speed penalty. The FIFO memory is one such device that is not affected by reset, since we are not interested in its initial values.

```
/**********************************************************/
// MODULE:          Synchronizing FIFO
//
// FILE NAME:       afifo_rtl.v
// VERSION:         1.1
// DATE:            January 1, 2003
// AUTHOR:          Bob Zeidman, Zeidman Consulting
//
// CODE TYPE:       Register Transfer Level
//
// DESCRIPTION:  This module defines a Synchronizing FIFO.
// The FIFO memory is implemented as a ring buffer. The read
// pointer points to the beginning of the buffer, while the
// write pointer points to the end of the buffer. Note that
// in this RTL version, the memory has one more location than
// the FIFO needs in order to calculate the FIFO count
// correctly.
//
/**********************************************************/

// DEFINES
`define DEL   1         // Clock-to-output delay. Zero
                        // time delays can be confusing
                        // and sometimes cause problems.

`define FIFO_DEPTH 15   // Depth of FIFO (number of bytes)
`define FIFO_HALF 8     // Half depth of FIFO
                        // (this avoids rounding errors)
`define FIFO_BITS 4     // Number of bits required to
                        // represent the FIFO size
`define FIFO_WIDTH 8    // Width of FIFO data

// TOP MODULE
module Afifo(
        reset_n,
        data_in,
        read_n,
        write_n,
        data_out,
        full,
        empty,
        half);

// PARAMETERS

// INPUTS
input                       reset_n;  // Active low reset
input [`FIFO_WIDTH-1:0]     data_in;  // Data input to FIFO
```

```
input                          read_n;   // Read FIFO (active low)
input                          write_n;  // Write FIFO (active low)
// OUTPUTS
output [`FIFO_WIDTH-1:0]       data_out; // FIFO data output
output                         full;     // FIFO is full
output                         empty;    // FIFO is empty
output                         half;     // FIFO is half full
                                         // or more

// INOUTS


// SIGNAL DECLARATIONS
wire                           reset_n;
wire [`FIFO_WIDTH-1:0]         data_in;
wire                           read_n;
wire                           write_n;
reg  [`FIFO_WIDTH-1:0]         data_out;
wire                           full;
wire                           empty;
wire                           half;


                               // The FIFO memory.
reg [`FIFO_WIDTH-1:0]          fifo_mem[0:`FIFO_DEPTH];
                               // How many locations in the FIFO
                               // are occupied?
wire [`FIFO_BITS-1:0]          counter;
                               // FIFO read pointer points to
                               // the location in the FIFO to
                               // read from next
reg [`FIFO_BITS-1:0]           rd_pointer;
                               // FIFO write pointer points to
                               // the location in the FIFO to
                               // write to next
reg [`FIFO_BITS-1:0]           wr_pointer;
// ASSIGN STATEMENTS
assign #`DEL counter = (wr_pointer >= rd_pointer) ?
                wr_pointer - rd_pointer :
                `FIFO_DEPTH + wr_pointer - rd_pointer + 1;
assign #`DEL full = (counter == `FIFO_DEPTH) ? 1'b1 : 1'b0;
assign #`DEL empty = (counter == 0) ? 1'b1 : 1'b0;
assign #`DEL half = (counter >= `FIFO_HALF) ? 1'b1 : 1'b0;

// MAIN CODE
```

```verilog
// This block contains all devices affected by the read
// and reset inputs
// Look at the falling edges of read_n and reset_n
always @(negedge reset_n or negedge read_n) begin
    if (~reset_n) begin
        // Reset the read pointer
        rd_pointer <= #`DEL `FIFO_BITS'b0;
    end
    else begin
        // Increment the read pointer
        // Check if the read pointer has gone beyond the
        // depth of the FIFO. If so, set it back to the
        // beginning of the FIFO
        // Note that >= is used instead of == because this
        // simplifies the logic
        // Note that the RTL version has a larger memory
        // than the Behavioral version to implement the
        // same size FIFO
        if (rd_pointer >= `FIFO_DEPTH)
            rd_pointer <= #`DEL `FIFO_BITS'b0;
        else
            rd_pointer <= #`DEL rd_pointer + 1;
    end
end

// This block contains all devices affected by the read
// input but not reset
// Look at the falling edge of read_n
always @(negedge read_n) begin
    // Check for FIFO underflow
    if (counter == 0) begin
        $display("\nERROR at time %0t:", $time);
        $display("FIFO Underflow\n");

        // Use $stop for debugging
        $stop;
    end

    // Output the data
    data_out <= #`DEL fifo_mem[rd_pointer];
end

// This block contains all devices affected by the write
// and reset inputs
// Look at the falling edges of write_n and reset_n
always @(negedge reset_n or negedge write_n) begin
    if (~reset_n) begin
```

```
            // Reset the write pointer
            wr_pointer <= #`DEL `FIFO_BITS'b0;
      end
      else begin
            // Increment the write pointer
            // Check if the write pointer has gone beyond the
            // depth of the FIFO. If so, set it back to the
            // beginning of the FIFO
            // Note that >= is used instead of == because this
            // simplifies the logic
            // Note that the RTL version has a larger memory
            // than the Behavioral version to implement the
            // same size FIFO
            if (wr_pointer >= `FIFO_DEPTH)
                wr_pointer <= #`DEL `FIFO_BITS'b0;
            else
                wr_pointer <= #`DEL wr_pointer + 1;

      end
end

// This block contains all devices affected by the write
// input but not reset
// Look at the falling edge of write_n
always @(negedge write_n) begin
      // Check for FIFO overflow
      if (counter >= `FIFO_DEPTH) begin
          $display("\nERROR at time %0t:", $time);
          $display("FIFO Overflow\n");

          // Use $stop for debugging
          $stop;
      end

      // Store the data
      fifo_mem[wr_pointer] <= #`DEL data_in;
end
endmodule      // Afifo
```

25.3 SIMULATION CODE

The simulation code for the FIFO is shown below. Note that the simulation uses the defined `fifo_depth` that is also used in the code for the hardware function. In a real design, it is useful to keep all of the defines in a separate file that can be included in each Verilog code file.

In that way, when the value of the define is changed, for any reason, that change is done once in the common file, and it affects each file where it is used in the code.

The simulation begins by writing quickly to the FIFO while reading slowly. This fills up the FIFO. Once the FIFO is filled, it changes the frequency of the reads and writes. Writing slowly and reading quickly, the FIFO empties and the simulation ends.

```
/***********************************************************/
// MODULE:          Synchronizing FIFO simulation
//
// FILE NAME:       afifo_sim.v
// VERSION:         1.1
// DATE:            January 1, 2003
// AUTHOR:          Bob Zeidman, Zeidman Consulting
//
// CODE TYPE:       Simulation
//
// DESCRIPTION:  This module provides stimuli for simulating
// a Synchronizing FIFO. It begins by writing quickly to the
// FIFO while reading slowly. This fills up the FIFO. Once
// the FIFO is filled, it changes the frequency of the reads
// and writes. Writing slowly and reading quickly, the FIFO
// empties and the simulation ends.
//
/***********************************************************/

// DEFINES
`define DEL   1         // Clock-to-output delay. Zero
                        // time delays can be confusing
                        // and sometimes cause problems.

`define FIFO_DEPTH 15   // Depth of FIFO (number of bytes)
`define FIFO_HALF 8     // Half depth of FIFO
                        // (this avoids rounding errors)
`define FIFO_WIDTH 8    // Width of FIFO data

// TOP MODULE
module afifo_sim();

// PARAMETERS

// INPUTS

// OUTPUTS

// INOUTS

// SIGNAL DECLARATIONS
```

```
reg                      clr_n;
reg    [`FIFO_WIDTH-1:0] in_data;
reg                      read_n;
reg                      write_n;
wire   [`FIFO_WIDTH-1:0] out_data;
wire                     full;
wire                     empty;
wire                     half;

integer                  fifo_count;   // Keep track of the number
                                       // of bytes in the FIFO
reg [`FIFO_WIDTH-1:0]    exp_data;     // Expected FIFO data
reg                      fast_read;      // Read at high frequency
reg                      fast_write;   // Write at high frequency
reg                      filled_flag;  // The FIFO has filled
                                       // at least once

// ASSIGN STATEMENTS

// MAIN CODE

// Instantiate the counter
Afifo afifo(
        .reset_n(clr_n),
        .data_in(in_data),
        .read_n(read_n),
        .write_n(write_n),
        .data_out(out_data),
        .full(full),
        .empty(empty),
        .half(half));

// Initialize inputs
initial begin
   in_data = 0;
   exp_data = 0;
   fifo_count = 0;
   read_n = 1;
   write_n = 1;
   filled_flag = 0;

   // Write quickly to the FIFO
   fast_write = 1;
   // Read slowly from the FIFO
   fast_read = 0;

   // Reset the FIFO
```

```
    clr_n = 0;
    #20 clr_n = 1;

    // Check that the status outputs are correct
    if (empty !== 1) begin
        $display("\nERROR at time %0t:", $time);
        $display("After reset, empty status not asserted\n");

        // Use $stop for debugging
        $stop;
    end
    if (full !== 0) begin
        $display("\nERROR at time %0t:", $time);
        $display("After reset, full status is asserted\n");

        // Use $stop for debugging
        $stop;
    end
    if (half !== 0) begin
        $display("\nERROR at time %0t:", $time);
        $display("After reset, half status is asserted\n");

        // Use $stop for debugging
        $stop;
    end

    // Start the action
    write_n <= #40 0;
    read_n <= #80 0;
end

// Simulate
// Write the FIFO
always @(negedge write_n) begin
    // Increment the count
    fifo_count = fifo_count + 1;

    // Bring write high
    #10 write_n = 1;

    // Set up the data for the next write
    #10 in_data = in_data + 1;

    // Do not write the FIFO if it is full
    wait (full === 0);

    // Set up the next falling edge of write
```

```verilog
   if (fast_write === 1)
      write_n <= #10 0;
   else
      write_n <= #30 0;
end

// Read the FIFO
always @(negedge read_n) begin
   // Decrement the count
   fifo_count = fifo_count - 1;

   // Set up the next falling edge of read
   if (fast_read === 1)
      #10;
   else
      #30;

   if (out_data !== exp_data) begin
      $display("\nERROR at time %0t:", $time);
      $display("   Expected data out = %h", exp_data);
      $display("   Actual data out   = %h\n", out_data);

      // Use $stop for debugging
      $stop;
   end

   // Bring read high to read the FIFO
   read_n = 1;

   // Increment the expected data
   exp_data = exp_data + 1;

   // Do not read the FIFO if it is empty
   wait (empty === 0);

   // Set up read for the next FIFO read
   read_n <= #20 0;
end

// Check all of the status signals with each change
// of fifo_count
always @(fifo_count) begin
   // Wait a moment to evaluate everything
   #`DEL;
   #`DEL
   #`DEL;
```

```
case (fifo_count)
    0: begin
        if ((empty !== 1) || (half !== 0) ||
                (full !== 0)) begin
            $display("\nERROR at time %0t:", $time);
            $display("    fifo_count = %h", fifo_count);
            $display("    empty = %b", empty);
            $display("    half  = %b", half);
            $display("    full  = %b\n", full);

            // Use $stop for debugging
            $stop;
        end

        if (filled_flag === 1) begin
            // The FIFO has filled and emptied
            $display("\nSimulation complete - no errors\n");
            $finish;
        end
    end
    `FIFO_HALF: begin
        if ((empty !== 0) || (half !== 1) ||
                (full !== 0)) begin
            $display("\nERROR at time %0t:", $time);
            $display("    fifo_count = %h", fifo_count);
            $display("    empty = %b", empty);
            $display("    half  = %b", half);
            $display("    full  = %b\n", full);

            // Use $stop for debugging
            $stop;
        end
    end
    `FIFO_DEPTH: begin
        if ((empty !== 0) || (half !== 1) ||
                (full !== 1)) begin
            $display("\nERROR at time %0t:", $time);
            $display("    fifo_count = %h", fifo_count);
            $display("    empty = %b", empty);
            $display("    half  = %b", half);
            $display("    full  = %b\n", full);

            // Use $stop for debugging
            $stop;
        end

        // The FIFO has filled, so set the flag
        filled_flag = 1;
```

```
            // Once the FIFO has filled, empty it
            // Write slowly to the FIFO
            fast_write = 0;
            // Read quickly from the FIFO
            fast_read = 1;
        end
        default: begin
            if ((empty !== 0) || (full !== 0)) begin
                $display("\nERROR at time %0t:", $time);
                $display("    fifo_count = %h", fifo_count);
                $display("    empty = %b", empty);
                $display("    half  = %b", half);
                $display("    full  = %b\n", full);

                // Use $stop for debugging
                $stop;
            end
            if (((fifo_count < `FIFO_HALF) &&
                    (half === 1)) ||
                ((fifo_count >= `FIFO_HALF) &&
                    (half === 0))) begin
                $display("\nERROR at time %0t:", $time);
                $display("    fifo_count = %h", fifo_count);
                $display("    empty = %b", empty);
                $display("    half  = %b", half);
                $display("    full  = %b\n", full);

                // Use $stop for debugging
                $stop;
            end
        end
    end
    endcase
end

endmodule    // afifo_sim
```

P A R T **7**

MEMORY CONTROLLERS

*T*he following chapters describe controllers for different kinds of memory devices. These controllers can be used to interface a generic microprocessor to a generic memory device. For specific processors and memory devices with unique features, they will need to be modified. However, they can be used as a framework for more sophisticated controllers.

The SRAM/ROM Controller

This chapter describes a simple state machine to control a Static Random Access Memory (SRAM) or Read Only Memory (ROM). Obviously, the difference between a ROM controller and an SRAM controller is simply that the ROM data cannot be written, so the write enable input to the chip should be tied high when the controller is used to access a ROM.

The connection between the processor, the SRAM, and the controller would typically look like the diagram in Figure 26-1 where the data and address buses are connected directly to the SRAM. It is the output enable (OE) and write enable (WE) to the SRAM, and the acknowledge (ACK) to the processor that must be derived from the processor address strobe (AS) and read/write (R/W) outputs. This is the job of the SRAM controller.

The state machine for the controller is shown in Figure 26-2. From the IDLE state, if there is a write input (AS asserted and RW low), the machine enters the WRITE state and does not leave until the counter reaches the necessary count. Similarly if there is a read input (AS asserted and RW high), the machine enters the READ state and does not leave until the

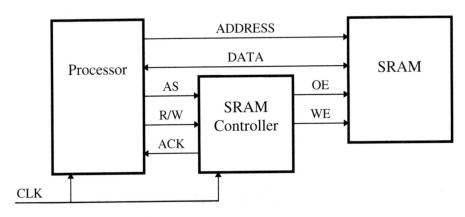

Figure 26-1 Connecting the Processor to the SRAM

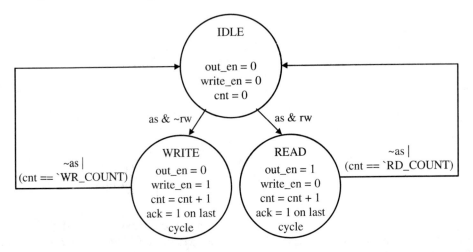

Figure 26-2 SRAM/ROM controller state machine.

counter reaches the necessary count. The number of cycles for writes and reads are controlled with `define` statements in the code. An acknowledge back to the processor is given on the last cycle of either access. Note that the access is aborted if the address strobe is deasserted by the processor before the access has completed.

26.1 BEHAVIORAL CODE

```
/***********************************************************/
// MODULE:       SRAM/ROM Controller
//
// FILE NAME:    sramcon_beh.v
```

```
// VERSION:      1.1
// DATE:         January 1, 2003
// AUTHOR:       Bob Zeidman, Zeidman Consulting
//
// CODE TYPE:    Behavioral Level
//
// DESCRIPTION:  This module implements a controller for an
// SRAM or ROM.
//
/************************************************************/

// DEFINES
`define DEL    1           // Clock-to-output delay. Zero
                           // time delays can be confusing
                           // and sometimes cause problems.
`define WR_COUNT 1         // Number of write cycles needed
`define RD_COUNT 4         // Number of read cycles needed
`define CNT_BITS 2         // Number of bits needed for the
                           // counter to count the cycles

// TOP MODULE
module sram_control(
     clock,
     reset_n,
     as_n,
     rw,
     out_en,
     write_en,
     ack);

// PARAMETERS
parameter[1:0]         // State machine states
   IDLE   = 0,
   WRITE  = 1,
   READ   = 2;

// INPUTS
input     clock;       // State machine clock
input     reset_n;     // Active low, synchronous reset
input     as_n;        // Active low address strobe
input     rw;          // Read/write command
                       // = 1 to read
                       // = 0 to write

// OUTPUTS
output    out_en;      // Output enable to memory
output    write_en;    // Write enable to memory
output    ack;         // Acknowledge signal to processor

// INOUTS
```

```verilog
// SIGNAL DECLARATIONS
wire        clock;
wire        reset_n;
wire        as_n;
wire        rw;
wire        out_en;
wire        write_en;
wire        ack;

reg  [1:0]                mem_state;        // State machine
reg  [`CNT_BITS-1:0]      cnt;            // Cycle counter

// ASSIGN STATEMENTS
                    // Create the outputs from the states
assign out_en = mem_state[1];
assign write_en = mem_state[0];

                    // Create the acknowledge combinatorially
assign #`DEL ack = ~as_n && ((~rw && (cnt == `WR_COUNT-1)) ||
                             ( rw && (cnt == `RD_COUNT-1)));

// MAIN CODE

// Look at the edge of reset
always @(reset_n) begin
    if (~reset_n) begin
        #`DEL assign mem_state = IDLE;
        assign cnt = `CNT_BITS'h0;
    end
    else begin
        #`DEL;
        deassign mem_state;
        deassign cnt;
    end
end

// Look at the rising edge of clock for state transitions
always @(posedge clock) begin
    case (mem_state)
        IDLE:  begin
            // Look for address strobe to begin the access
            if (~as_n) begin
                if (rw) begin
                    // This is a read access
                    mem_state <= #`DEL READ;
                end
                else begin
                    // This is a write access
                    mem_state <= #`DEL WRITE;
                end
            end
```

```
        end
    WRITE: begin
        // If we have reached the final cycle count
        // for the access, the access is finished.
        // If the address strobe has been deasserted,
        // the access is aborted
        if ((cnt == `WR_COUNT-1) || as_n) begin
            mem_state <= #`DEL IDLE;
            cnt <= #`DEL `CNT_BITS'h0;
        end
        else
            cnt <= #`DEL cnt + 1;
    end
    READ: begin
        // If we have reached the final cycle count
        // for the access, the access is finished.
        // If the address strobe has been deasserted,
        // the access is aborted
        if ((cnt == `RD_COUNT-1) || as_n) begin
            mem_state <= #`DEL IDLE;
            cnt <= #`DEL `CNT_BITS'h0;
        end
        else
            cnt <= #`DEL cnt + 1;
    end
    endcase
end
endmodule    // sram_control
```

26.2 RTL CODE

The following RTL code implements the SRAM/ROM controller. Note that the acknowledge output is generated using an *assign* statement. This means that the resulting logic will be combinatorial, based on the state machine outputs. The advantage to this is that the state machine can be designed for single cycle reads or writes because it can assert the acknowledge output on the same cycle that it receives the address strobe input. The disadvantage is that the clock-to-output timing of the acknowledge will not be optimal. If there are no single cycle accesses, the acknowledge can be generated from a flip-flop instead, which will reduce the clock-to-output time and may then decrease the required clock cycle time.

```
/***********************************************************/
// MODULE:          SRAM/ROM Controller
//
// FILE NAME:       sramcon_rtl.v
// VERSION:         1.1
// DATE:            January 1, 2003
```

```
// AUTHOR:        Bob Zeidman, Zeidman Consulting
//
// CODE TYPE:     Register Transfer Level
//
// DESCRIPTION:  This module implements a controller for an
// SRAM or ROM.
//
/*********************************************************/

// DEFINES
`define DEL   1          // Clock-to-output delay. Zero
                         // time delays can be confusing
                         // and sometimes cause problems.
`define WR_COUNT 1       // Number of write cycles needed
`define RD_COUNT 4       // Number of read cycles needed
`define CNT_BITS 2       // Number of bits needed for the
                         // counter to count the cycles
// TOP MODULE
module sram_control(
      clock,
      reset_n,
      as_n,
      rw,
      out_en,
      write_en,
      ack);

// PARAMETERS
parameter[1:0]        // State machine states
    IDLE  = 0,
    WRITE = 1,
    READ  = 2;

// INPUTS
input         clock;    // State machine clock
input         reset_n;    // Active low, synchronous reset
input         as_n;     // Active low address strobe
input         rw;       // Read/write command
                        // = 1 to read
                        // = 0 to write

// OUTPUTS
output        out_en;   // Output enable to memory
output        write_en; // Write enable to memory
output        ack;      // Acknowledge signal to processor

// INOUTS

// SIGNAL DECLARATIONS
wire          clock;
wire          reset_n;
```

```
wire        as_n;
wire        rw;
wire        out_en;
wire        write_en;
wire        ack;

reg  [1:0]              mem_state;        // Synthesis state_machine
reg  [`CNT_BITS-1:0]    cnt;              // Cycle counter

// ASSIGN STATEMENTS
                    // Create the outputs from the states
assign out_en = mem_state[1];
assign write_en = mem_state[0];

                    // Create the acknowledge combinatorially
assign #`DEL ack = ~as_n && ((~rw && (cnt == `WR_COUNT-1)) ||
                             ( rw && (cnt == `RD_COUNT-1)));

// MAIN CODE

// Look at the rising edge of clock for state transitions
always @(posedge clock or negedge reset_n) begin
    if (~reset_n) begin
        mem_state <= #`DEL IDLE;
        cnt <= #`DEL `CNT_BITS'h0;
    end
    else begin
                        // Use parallel_case directive
                        // to show that all states are
                        // mutually exclusive
                        // Use full_case directove
                        // to show that any other states
                        // are don't cares
        case (mem_state) // synthesis parallel_case full_case
            IDLE: begin
                // Look for address strobe to begin the access
                if (~as_n) begin
                    if (rw) begin
                        // This is a read access
                        mem_state <= #`DEL READ;
                    end
                    else begin
                        // This is a write access
                        mem_state <= #`DEL WRITE;
                    end
                end
            end
            WRITE: begin
                // If we have reached the final cycle count
                // for the access, the access is finished.
                // If the address strobe has been deasserted,
```

```
                    // the access is aborted
                    if ((cnt == `WR_COUNT-1) || as_n) begin
                        mem_state <= #`DEL IDLE;
                        cnt <= #`DEL `CNT_BITS'h0;
                    end
                    else
                        cnt <= #`DEL cnt + 1;
                end
                READ: begin
                    // If we have reached the final cycle count
                    // for the access, the access is finished.
                    // If the address strobe has been deasserted,
                    // the access is aborted
                    if ((cnt == `RD_COUNT-1) || as_n) begin
                        mem_state <= #`DEL IDLE;
                        cnt <= #`DEL `CNT_BITS'h0;
                    end
                    else
                        cnt <= #`DEL cnt + 1;
                end
            endcase
        end
end
endmodule     // sram_control
```

26.3 SIMULATION CODE

The simulation code uses a simple SRAM memory model. The simulation instantiates the memory controller and the SRAM. First it asserts the reset signal and checks that outputs are reset to the correct values. Then it writes a series of data words to each successive location in memory. It then reads back those locations and checks that the data words match those that were previously written. Note that tasks are used for read and write operations. Since a number of these operations are performed, using tasks greatly reduces the amount of code to be written.

Designing the memory posed some problems, because it is necessary that the memory check that the read and write signals have roughly the correct pulse widths. In this book, I have not included specific timing information, because that is not the purpose of this book. Timing checking is very involved, and depends heavily of the particular technologies being used. But the main function of the SRAM controller is to provide read and write signals that have the correct timing. So if we do not test the timing of these signals, we are not really testing the SRAM controller.

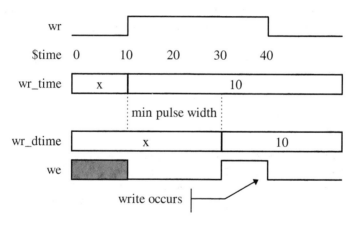

Figure 26-3 SRAM Controller Passes Timing Test.

To see how these timing parameters are tested, look at how a write access is modeled in the `memory` submodule, shown graphically in Figures 26-3 and 26-4. We look first at the leading edge of the `wr` signal. At this time, we save the value of the current value of `$time` in a variable called `wr_time`. An assign statement causes another signal, `wr_dtime`, to take on the same value as `wr_time`, but one minimum write pulse length later in the simulation. When the delayed signal, `wr_dtime`, changes, we compare it to the original variable `wr_time`. If they are equal, then there have been no other leading edges and the write pulse has met the minimum duration requirement. In that case, we assert a write enable signal, `we`, that means that the write access is enabled and will occur on the falling edge of the `wr` input. If the values are different, then another leading edge occurred in the meantime, that means that the `wr` input was deasserted at some point and did not meet the minimum pulse width. In this case, a write enable signal is not asserted, and the write cannot take place. The read access uses the same mechanism to test the minimum pulse width timing.

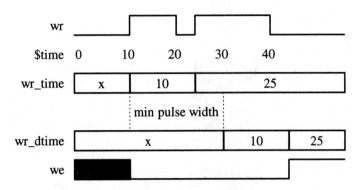

Figure 26-4 SRAM Controller Fails Timing Test.

```
/**********************************************************/
// MODULE:          SRAM/ROM Controller simulation
//
// FILE NAME:       sramcon_sim.v
// VERSION:         1.1
// DATE:            January 1, 2003
// AUTHOR:          Bob Zeidman, Zeidman Consulting
//
// CODE TYPE:       Simulation
//
// DESCRIPTION: This module provides stimuli for simulating
// an SRAM/ROM memory controller. It uses an SRAM model and
// performs a number of writes and reads to the model,
// including back-to-back reads and writes. It also checks
// the reset function.
//
/**********************************************************/

// DEFINES
`define DEL   1             // Clock-to-output delay. Zero
                            // time delays can be confusing
                            // and sometimes cause problems.
`define WR_COUNT 1          // Number of write cycles needed
`define RD_COUNT 4          // Number of read cycles needed
`define CLK_HPERIOD 50      // Clock half period
                            // Clock period
`define CLK_PERIOD CLK_HPERIOD*2
`define WR_PERIOD 70        // Write access time of memory
                            // Should be between
                            // (WR_COUNT-1)*CLK_PERIOD and
                            // WR_COUNT*CLK_PERIOD
```

```
`define RD_PERIOD 350      // Read access time of memory
                           // (RD_COUNT-1)*CLK_PERIOD and
                           // RD_COUNT*CLK_PERIOD
`define DWIDTH 8           // Data bus width
`define ADEPTH 256         // Memory depth
`define AWIDTH 8           // Number of bits needed to
                           // address memory

// TOP MODULE
module sramcon_sim();

// PARAMETERS

// INPUTS

// OUTPUTS

// INOUTS

// SIGNAL DECLARATIONS
reg           clock;
reg           reset;
reg           as;
reg           rw;
wire          out_en;
wire          write_en;
wire          ack;

reg  [`DWIDTH-1:0]  data_out;    // Data out of processor
wire [`DWIDTH-1:0]  data;        // Data bus
reg  [`AWIDTH-1:0]  address;     // Address bus

integer             cyc_count;   // Count cycles to determine
                                 // if the access has taken
                                 // too long
reg  [`AWIDTH:0]    addr_count;  // Address counter
reg  [`DWIDTH-1:0]  data_patt;   // Data pattern holder



// ASSIGN STATEMENTS
assign data = (as & rw) ? 8'hzz : data_out;

// MAIN CODE

// Instantiate the controller
sram_control Sram_control(
      .clock(clock),
      .reset_n(~reset),
      .as_n(~as),
```

```verilog
        .rw(rw),
        .out_en(out_en),
    ·   .write_en(write_en),
        .ack(ack));

// Instantiate an SRAM
sram Sram(
        .rd(out_en),
        .wr(write_en),
        .address(address),
        .data(data));

// Generate the clock
always #`CLK_HPERIOD clock = ~clock;

// Simulate
initial begin
    // Initialize inputs
    clock = 1;
    reset = 0;
    as = 0;

    // Test the reset signal
    #`DEL;
    // Assert the reset signal
    reset = 1;

    // Wait for the outputs to change asynchronously
    #`DEL
    #`DEL
    // Test outputs
    if ((out_en === 1'b0) && (write_en === 1'b0) &&
       (ack === 1'b0))
        $display ("Reset is working");
    else begin
        $display("\nERROR at time %0t:", $time);
        $display("Reset is not working");
        $display("    out_en   = %b", out_en);
        $display("    write_en = %b", write_en);
        $display("    ack      = %b\n", ack);

        // Use $stop for debugging
        $stop;
    end

    // Deassert the reset signal
    reset = 0;
```

```
    // Initialize the address counter
    addr_count = 0;

    // Initialize the data pattern to be written
    data_patt = `DWIDTH'h1;

    // Write a series of values to memory
    while (addr_count[`AWIDTH] === 0) begin
        // Write to memory
        writemem(addr_count[`AWIDTH-1:0], data_patt);

        // Increment the address counter
        addr_count <= addr_count + 1;

        // Shift the data pattern
        data_patt <= (data_patt << 1);
        data_patt[0] <= data_patt[`DWIDTH-1];

        // Wait for the data pattern to change
        #`DEL;
    end

    // Initialize the address counter
    addr_count = 0;

    // Initialize the data pattern to be read
    data_patt = `DWIDTH'h1;

    // Verify the values that were written
    while (addr_count[`AWIDTH] === 0) begin
        // Read from memory
        readmem(addr_count[`AWIDTH-1:0], data_patt);

        // Increment the address counter
        addr_count <= addr_count + 1;

        // Shift the data pattern
        data_patt <= (data_patt << 1);
        data_patt[0] <= data_patt[`DWIDTH-1];

        // Wait for the data pattern to change
        #`DEL;
    end

    $display("\nSimulation complete - no errors\n");
    $finish;
end
```

```
// Check the number of cycles for each access
always @(posedge clock) begin
    // Check whether an access is taking too long
    if (cyc_count > (`RD_COUNT + `WR_COUNT)) begin
        $display("\nERROR at time %0t:", $time);
        if (rw)
            $display("Read access took to long\n");
        else
            $display("Write access took to long\n");

        // Use $stop for debugging
        $stop;
    end
end

// TASKS
// Write data to memory
task writemem;

// INPUTS
input [`AWIDTH-1:0] write_addr;  // Memory address
input [`DWIDTH-1:0] write_data;  // Data to write to memory

// OUTPUTS

// INOUTS

// TASK CODE
begin
    cyc_count = 0;      // Initialize the cycle count

    // Wait for the rising clock edge
    @(posedge clock);
    rw <= 0;            // Set up a write access
    as <= 1;            // Assert address strobe
    address <= write_addr; // Set up the address
    data_out <= write_data; // Set up the data

    // Wait for the acknowledge
    @(posedge clock);
    while (~ack) begin
        // Increment the cycle count
        cyc_count = cyc_count + 1;

        @(posedge clock);
    end
```

```
    as <= 0;              // Deassert address strobe
    @(posedge clock);    // Wait one clock cycle

end
endtask        // writemem

// Read data from memory and check its value
task readmem;

// INPUTS
input [`AWIDTH-1:0] read_addr;   // Memory address
input [`DWIDTH-1:0] expected;    // Expected read data

// OUTPUTS

// INOUTS

// TASK CODE
begin
    cyc_count = 0;               // Initialize the cycle count

    // Wait for the rising clock edge
    @(posedge clock);
    rw <= 1;                     // Set up a read access
    as <= 1;                     // Assert address strobe
    address <= read_addr;        // Set up the address

    // Wait for the acknowledge
    @(posedge clock);
    while (~ack) begin
        // Increment the cycle count
        cyc_count = cyc_count + 1;

        @(posedge clock);
    end

    // Did we find the expected data?
    if (data !== expected) begin
        $display("\nERROR at time %0t:", $time);
        $display("Controller is not working");
        $display("    data written = %h", expected);
        $display("    data read    = %h\n", data);

        // Use $stop for debugging
        $stop;
    end
```

```
    as <= 0;              // Deassert address strobe
    @(posedge clock);    // Wait one clock cycle

end
endtask          // readmem

endmodule        // sramcon_sim

// SUBMODULE
// SRAM memory model
module sram(
        data,
        address,
        rd,
        wr);

// INPUTS
input [`AWIDTH-1:0] address;   // Memory address
input                  rd;     // Read input
input             wr;          // Write input

// OUTPUTS

// INOUTS
inout [`DWIDTH-1:0] data;      // Data lines

// SIGNAL DECLARATIONS
wire [`AWIDTH-1:0]  address;
wire                  rd;
wire                  wr;
wire [`DWIDTH-1:0]  data;
reg  [`DWIDTH-1:0]  mem[`ADEPTH-1:0];   // Stored data

time          rd_time;      // Time of the most recent change
                            // of the rd input
wire [63:0]  rd_dtime;      // rd_time delayed by `RD_PERIOD
                            // time units

time          wr_time;      // Time of the most recent change
                            // of the wr input
wire [63:0]  wr_dtime;      // wr_time delayed by `WD_PERIOD
                            // time units

reg      oe;                // Output enable - this signal is
                            // asserted after the minimum rd
                            // pulse time
```

```
reg        we;               // Write enable - this signal is
                             // asserted after the minimum wr
                             // pulse time

// PARAMETERS

// ASSIGN STATEMENTS
assign #`RD_PERIOD rd_dtime = rd_time;
assign #`WR_PERIOD wr_dtime = wr_time;

// Output the data if the rd signal is asserted
// and the oe is asserted, which means that
// we met the minimum rd pulse time
assign data = (oe & rd) ? mem[address] : 8'hzz;

// MAIN CODE
initial begin
   oe = 0;
   we = 0;
   wr_time = 0;
   rd_time = 0;
end

// Look for the beginning of the write access
always @(posedge wr) begin
   wr_time = $time;         // Save the time of the most
                            // recent rising edge of wr
   we = 0;
end

always @(wr_dtime) begin
   if (wr_time === wr_dtime) begin
      // If this is true, the last rising edge of wr
      // was exactly `WR_PERIOD time units ago, so
      // assert the write enable
      we = 1;
   end
end

// Look for the end of the write access
always @(negedge wr) begin
   if (we) begin
      // Write the data if the write enable is asserted
      // which means we met the minimum wr pulse time
      mem[address] <= data;
   end
   we = 0;               // Turn off the write enable
end
```

```
// Look for the beginning of the read access
always @(posedge rd) begin
    rd_time = $time;                 // Save the time of the most
                                     // recent rising edge of rd

    oe = 0;
end

always @(rd_dtime) begin
    if (rd_time === rd_dtime) begin
        // If this is true, the last rising edge of rd
        // was exactly `RD_PERIOD time units ago, so
        // assert the output enable
        oe = 1;
    end
end

// Look for the end of the read access
always @(negedge rd) begin
    oe = 0;              // Turn off the output enable
end
endmodule     // sram
```

counter reaches the necessary count. The number of cycles for writes and reads are controlled with `define` statements in the code. An acknowledge back to the processor is given on the last cycle of either access. Note that the access is aborted if the address strobe is deasserted by the processor before the access has completed.

counter reaches the necessary count. The number of cycles for writes and reads are controlled with `define` statements in the code. An acknowledge back to the processor is given on the last cycle of either access. Note that the access is aborted if the address strobe is deasserted by the processor before the access has completed.

The Synchronous SRAM Controller

A relatively new type of SRAM has been developed in order to interface to a very fast processor. This type of SRAM is called a Synchronous SRAM (SSRAM) because the inputs and outputs are synchronized to a single clock. This makes it especially good for use with a Reduced Instruction Set Computer (RISC) type processor because these processors tend to execute one or more instructions per clock cycle. When burst accesses of the SSRAM are required, the SSRAM can execute a read or write each cycle after an initial one or two cycle delay. This allows the processor to execute instructions at full speed, without the need to wait for a memory access to complete, if the processor allows pipelined accesses. During a pipelined access, the processor outputs the address and control signals for an access before the previous access has completed.

In this chapter I describe a state machine to interface an SSRAM to a pipelined RISC processor. There are two basic types of SSRAMs, called "flow-through" and "pipelined." A flow-through SSRAM has one cycle delay between the address/control cycle of an access and the data cycle of an access. A pipelined SSRAM has two cycles of delay between the address/control cycle of an access and the data cycle of an access. The read and write timing for a flow-through SSRAM is shown in Figure 27-1. The read and write timing for a pipelined SSRAM is shown in Figure 27-2.

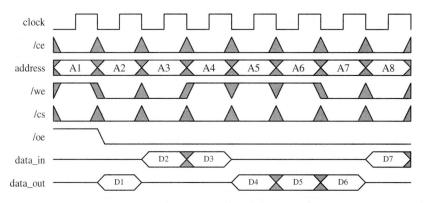

Figure 27-1 Flowthrough SSRAM read and write cycle waveforms.

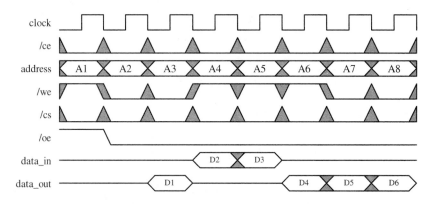

Figure 27-2 Pipelined SSRAM read and write cycle waveforms.

The connection between the processor, the SSRAM, and the controller looks like the diagram in Figure 27-3 where the data and address buses and the write enable (WE) signal are connected directly to the SRAM. The output enable (OE) of the SSRAM is tied to a logic one and the next access (NA) input to the processor is also tied to a logic one. The next access input specifies that the processor can specify the address and control inputs for the next access even though the current access has not yet completed. The SSRAM controller simply takes the address strobe (AS) signal from the processor and uses it to generate the acknowledge signal (ACK) back to the processor.

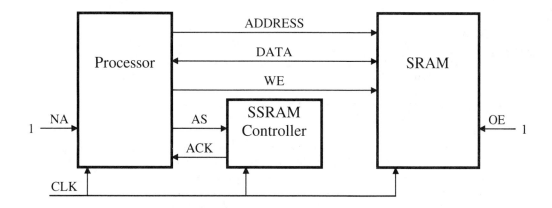

Figure 27-3 Connecting the processor to the SSRAM.

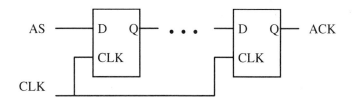

Figure 27-4 SSRAM controller.

The controller turns out to be simply a shift register which delays the address strobe by a number of clock cycles to generate the acknowledge, as shown in Figure 27-4. For a flow-through SSRAM, the shift register consists of one flip-flop, which delays one clock cycle. For a pipelined SSRAM., the shift register consists of two flip-flops, which delays two clock cycles. Remember that this simple controller works for a hypothetical RISC processor. For a Complex Instruction Set Computer (CISC), the interface logic is typically more complex.

27.1 BEHAVIORAL CODE

```
/*****************************************************************/
// MODULE:          Synchronous SRAM Controller
//
// FILE NAME:       ssramcon_beh.v
// VERSION:         1.1
// DATE:            January 1, 2003
// AUTHOR:          Bob Zeidman, Zeidman Consulting
```

```
//
// CODE TYPE:       Behavioral Level`
//
// DESCRIPTION:  This module implements a controller for an
// Synchronous SRAM.
//
/**********************************************************/

// DEFINES
`define DEL   1      // Clock-to-output delay. Zero
                     // time delays can be confusing
                     // and sometimes cause problems.
`define WAIT 2       // Number of wait cycles needed
                     // Flow-through SSRAMs require
                     // 1 wait cycles
                     // Pipelined SSRAMs require
                     // 2 wait cycle

// TOP MODULE
module ssram_control(
     clock,
     reset,
     as,
     ack);

// PARAMETERS

// INPUTS
input           clock;       // State machine clock
input           reset;       // Reset
input           as;          // Address strobe

// OUTPUTS
output          ack;         // Acknowledge to processor

// INOUTS

// SIGNAL DECLARATIONS
wire            clock;
wire            reset;
wire            as;
wire            ack;

reg  [`WAIT-1:0] shift_reg;       // Shift register

// ASSIGN STATEMENTS
// Create the acknowledge signal
```

```
assign ack = shift_reg[`WAIT-1];

// MAIN CODE

// Look at the edge of reset
always @(reset) begin
    if (reset) begin
        #`DEL assign shift_reg = 0;
    end
    else begin
        #`DEL deassign shift_reg;
    end
end

// Look at the rising edge of clock for state transitions
always @(posedge clock) begin
    shift_reg <= #`DEL shift_reg << 1;
    shift_reg[0] <= #`DEL as;
end
endmodule     // ssram_control
```

27.2 RTL CODE

```
/*********************************************************/
// MODULE:        Synchronous SRAM Controller
//
// FILE NAME:     ssramcon_rtl.v
// VERSION:       1.1
// DATE:          January 1, 2003
// AUTHOR:        Bob Zeidman, Zeidman Consulting
//
// CODE TYPE:     Register Transfer Level
//
// DESCRIPTION:  This module implements a controller for an
// Synchronous SRAM.
//
/*********************************************************/

// DEFINES
`define DEL   1     // Clock-to-output delay. Zero
                    // time delays can be confusing
                    // and sometimes cause problems.
`define WAIT 2      // Number of wait cycles needed
                    // Flow-through SSRAMs require
```

```
                              // 1 wait cycles
                              // Pipelined SSRAMs require
                              // 2 wait cycle

// TOP MODULE
module ssram_control(
        clock,
        reset,
        as,
        ack);

// PARAMETERS

// INPUTS
input           clock;        // State machine clock
input           reset;        // Reset
input           as;           // Address strobe

// OUTPUTS
output          ack;          // Acknowledge to processor

// INOUTS

// SIGNAL DECLARATIONS
wire            clock;
wire            reset;
wire            as;
wire            ack;

reg  [`WAIT-1:0] shift_reg;        // Shift register

// ASSIGN STATEMENTS
// Create the acknowledge signal
assign ack = shift_reg[`WAIT-1];

// MAIN CODE

// Look at the rising edge of clock for state transitions
always @(posedge clock or posedge reset) begin
   if (reset) begin
       shift_reg <= #`DEL 0;
   end
   else begin
       shift_reg <= #`DEL shift_reg << 1;
       shift_reg[0] <= #`DEL as;
   end
end
endmodule    // ssram_control
```

27.3 SIMULATION CODE

The simulation code uses a simple SSRAM memory model. The simulation instantiates the memory controller and the SRAM. First it asserts the reset signal and checks that ACK output is reset to the correct value. Then it writes a series of data words to each successive location in memory. It then reads back those locations and checks that the data words match those that were previously written. As before, tasks are used for read and write operations in order to reduce the amount of code that needs to be written. It is assumed that the processor can pipeline reads and writes in the same way that the SSRAM can, so that no wait states are necessary. Otherwise, wait states would need to be added, which reduces the advantage of a synchronous SRAM over a normal asynchronous one.

```
/*************************************************************/
// MODULE:          SSRAM Controller simulation
//
// FILE NAME:       ssramcon_sim.v
// VERSION:         1.1
// DATE:            January 1, 2003
// AUTHOR:          Bob Zeidman, Zeidman Consulting
//
// CODE TYPE:       Simulation
//
// DESCRIPTION:  This module provides stimuli for simulating
// an SSRAM memory controller. It uses an SSRAM model and
// performs a number of writes and reads to the model,
// including back-to-back reads and writes. It also checks
// the reset function.
//
/*************************************************************/

// DEFINES
`define DEL   1              // Clock-to-output delay. Zero
                             // time delays can be confusing
                             // and sometimes cause problems.
`define WAIT 2               // Number of wait cycles needed
                             // Flow-through SSRAMs require
                             // 1 wait cycles
                             // Pipelined SSRAMs require
                             // 2 wait cycle
`define DWIDTH 8             // Data bus width
`define ADEPTH 256           // Memory depth
`define AWIDTH 8             // Number of bits needed to
                             // address memory
`define CYCLE_TIME 100       // System clock cycle time

// TOP MODULE
```

```
module ssramcon_sim();

// PARAMETERS

// INPUTS

// OUTPUTS

// INOUTS

// SIGNAL DECLARATIONS
reg                      clock;
reg                      reset;
reg                      as;
reg                      we;
reg                      oe;
reg                      ce;
reg                      cs;
wire                     ack;

reg   [`DWIDTH-1:0]  data_out;      // Data written by processor
reg   [`DWIDTH-1:0]  data_exp;      // Data expected to be read
                                    // from memory
wire  [`DWIDTH-1:0]  data;          // Data bus
reg   [`AWIDTH-1:0]  address;       // Address bus
reg   [`AWIDTH-1:0]  addr_count;    // Address counter

reg   [`DWIDTH-1:0]  data_patt;      // Data pattern holder
reg   [`WAIT-1:0]    as_shift;      // Shift register for as
reg   [`WAIT-1:0]    we_shift;      // Shift register for we

integer              fin_flag;      // Tells the simulation
                                    // when we're finished


// ASSIGN STATEMENTS
// The data bus direction is determined by the address
// strobe and write enable signals from previous cycles.
assign data = (as_shift[`WAIT-1] & we_shift[`WAIT-1]) ?
              data_out : 8'hzz;

// MAIN CODE

// Instantiate the controller
ssram_control Ssram_control(
        .clock(clock),
        .reset(reset),
        .as(as),
        .ack(ack));
```

```verilog
// Instantiate an SSRAM
ssram Ssram(
        .clock(clock),
        .oe(oe),
        .we(we),
        .ce(ce),
        .cs(cs),
        .address(address),
        .data(data));

// Generate the clock
always #(`CYCLE_TIME/2) clock = ~clock;

// Simulate
initial begin
    // Initialize inputs
    clock = 1;
    reset = 0;
    as_shift = 0;
    we_shift = 0;
    as = 0;
    cs = 1;
    oe = 1;
    ce = 1;

    // Test the reset signal
    #`DEL;
    // Assert the reset signal
    reset = 1;

    // Wait for the outputs to change asynchronously
    #`DEL
    #`DEL
    // Test outputs
    if (ack === 1'b0)
        $display ("Reset is working");
    else begin
        $display("\nERROR at time %0t:", $time);
        $display("Reset is not working");
        $display("    ack  = %b\n", ack);

        // Use $stop for debugging
        $stop;
    end

    // Deassert the reset signal
    reset = 0;
```

```
// Initialize the address counter
addr_count = 0;

// Initialize the data pattern to be written
data_patt = `DWIDTH'h1;

// Write a series of values to memory
while (&addr_count === 1'b0) begin
    // Write to memory
    writemem(addr_count, data_patt);

    // Increment the address counter
    addr_count = addr_count + 1;

    // Shift the data pattern
    data_patt <= (data_patt << 1);
    data_patt[0] <= data_patt[`DWIDTH-1];

    // Wait for the data pattern to change
    #`DEL;
end
// Write once more to memory
writemem(addr_count, data_patt);

// Wait for the access to complete
#`DEL;

// Initialize the address counter
addr_count = 0;

// Initialize the data pattern to be read
data_patt = `DWIDTH'h1;

// Verify the values that were written
while (&addr_count === 1'b0) begin
    // Read from memory
    readmem(addr_count, data_patt);

    // Increment the address counter
    addr_count = addr_count + 1;

    // Shift the data pattern
    data_patt <= (data_patt << 1);
    data_patt[0] <= data_patt[`DWIDTH-1];

    // Wait for the data pattern to change
    #`DEL;
end
```

```
    // Read once more from memory
    readmem(addr_count, data_patt);

    // Set the finish flag
    fin_flag = `WAIT;

    // Turn off chip select
    @(posedge clock);
    cs <= #`DEL 0;
end

// Shift the address strobe and write enable signals
// so that we can use them after the appropriate cycle
// delay
always @(posedge clock) begin
    // Shift the address strobe signal into
    // the address strobe shift register
    as_shift <= #`DEL as_shift << 1;
    as_shift[0] <= #`DEL as;

    // Shift the write enable signal into
    // the write enable shift register
    we_shift <= #`DEL we_shift << 1;
    we_shift[0] <= #`DEL we;

    // Look for an acknowledge during the correct cycle
    if (as_shift[`WAIT-1] & ~ack) begin
        $display("\nERROR at time %0t:", $time);
        $display("Did not receive the expected acknowledge\n");

        // Use $stop for debugging
        $stop;
    end
    else if (~as_shift[`WAIT-1] & ack) begin
        $display("\nERROR at time %0t:", $time);
        $display("Received an unexpected acknowledge\n");

        // Use $stop for debugging
        $stop;
    end
end

// Perform any necessary signal checking
always @(data_exp) begin
    // Did we find the expected data?
    if (data !== data_exp) begin
        $display("\nERROR at time %0t:", $time);
```

```
        $display("Controller is not working");
        $display("    data written = %h", data_exp);
        $display("    data read    = %h\n", data);

        // Use $stop for debugging
        $stop;
    end

    // Are we done checking all reads?
    if (fin_flag === 0) begin
        $display("\nSimulation complete - no errors\n");
        $finish;
    end
    fin_flag = fin_flag - 1;
end

// TASKS
// Write data to memory
task writemem;

// INPUTS
input [`AWIDTH-1:0] write_addr;   // Memory address
input [`DWIDTH-1:0] write_data;   // Data to write to memory

// OUTPUTS

// INOUTS

// TASK CODE
begin
    // Wait for the rising clock edge
    @(posedge clock);
    we <= #`DEL 1;                  // Set up a write access
    as <= #`DEL 1;                  // Assert address strobe
    address <= #`DEL write_addr;    // Set up the address
                                    // Set up the data to change
                                    // at the correct cycle
    data_out <= #(`CYCLE_TIME * `WAIT + `DEL) write_data;
end
endtask      // writemem

// Read data from memory and check its value
task readmem;

// INPUTS
input [`AWIDTH-1:0] read_addr;   // Memory address
input [`DWIDTH-1:0] read_data;   // Expected read data
```

```
// OUTPUTS

// INOUTS

// TASK CODE
begin
    // Wait for the rising clock edge
    @(posedge clock);
    we <= #`DEL 0;                   // Set up a read access
    as <= #`DEL 1;                   // Assert address strobe
    address <= #`DEL read_addr;      // Set up the address

    // Set up the expected data
    data_exp <= #(`CYCLE_TIME * `WAIT + 2*`DEL) read_data;
end
endtask        // readmem

endmodule          // ssramcon_sim

// SUBMODULE
// SSRAM memory model
module ssram(
      clock,
      data,
      address,
      oe,
      ce,
      cs,
      we);

// INPUTS
input                 clock;     // Clock
input [`AWIDTH-1:0]   address;   // Memory address
input                 oe;        // Output enable
input                 we;        // Write enable
input                 ce;        // Clock enable
input                 cs;        // Chip select

// OUTPUTS

// INOUTS
inout [`DWIDTH-1:0]   data;      // Data lines

// SIGNAL DECLARATIONS
wire                  clock;
wire [`AWIDTH-1:0]    address;
wire                  oe;
wire                  we;
```

```
wire               ce;
wire               cs;
wire [`DWIDTH-1:0] data;

reg  [`DWIDTH-1:0] mem[`ADEPTH-1:0];   // Stored data

                                // Shift register for storing
                                // addresses for pipelined accesses
reg  [`AWIDTH-1:0] addr_shift[`WAIT-1:0];
                                // Shift register for storing write
                                // enable for pipelined accesses
reg  [`WAIT-1:0]   we_shift;
                                // Shift register for storing chip
                                // select for pipelined accesses
reg  [`WAIT-1:0]   cs_shift;

reg                i;          // Temporary variable

// PARAMETERS

// ASSIGN STATEMENTS

// Output the data if the SSRAM is enabled, the clock is
// enabled, oe is asserted, and the current access is a read
assign data = ce & oe & cs_shift[`WAIT-1] &
       ~we_shift[`WAIT-1] ?mem[addr_shift[`WAIT-1]] : 8'hzz;

// MAIN CODE

// Initialize chip select shift register
initial begin
   cs_shift = 0;
end

// Look at the rising edge of clock
always @(posedge clock) begin
   // Don't do anything unless the clock is enabled
   if (ce) begin
      // Store the address and control signals
      // in their respective shift registers and shift them
      cs_shift <= #`DEL cs_shift << 1;
      cs_shift[0] <= #`DEL cs;
      we_shift <= #`DEL we_shift << 1;
      we_shift[0] <= #`DEL we;

      for (i = `WAIT-1; i > 0; i = i - 1) begin
         addr_shift[i] <= #`DEL addr_shift[i-1];
      end
```

```
      addr_shift[0] <= #`DEL address;

      // If there is a chip select and write enable,
      // write the data
      if (cs_shift[`WAIT-1] & we_shift[`WAIT-1]) begin
            mem[addr_shift[`WAIT-1]] <= #`DEL data;
      end
   end
end
endmodule    // sram
```

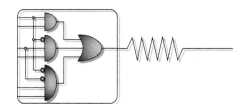

The DRAM Controller

*T*his chapter describes a state machine to control a Dynamic Random Access Memory. The connection between the processor, the DRAM, and the controller typically looks like the diagram in Figure 28-1, where the data bus is connected directly to the DRAM. The Row Address Strobe (RAS) and Column Address Strobe (CAS) to the DRAM, and the Acknowledge input to the processor (ACK) are derived from the processor Address Strobe (AS) by the controller. Also, the controller must include a periodic refresh of the DRAM using the RAS and CAS inputs. Note that the Read/Write signal goes through the controller. In many cases, this signal can be connected directly to the DRAM. However, some DRAMs require that the Read/Write signal be high during a refresh cycle. In that case, the controller must keep the Read/Write signal to the DRAM high during a refresh cycle, rather than take it directly from the processor.

Note that the following discussion assumes that the signals from the processor are synchronized to the clock signal. If this is not so, then the address strobe should be synchronized to the clock by sending it through at least one flip-flop, before it is used in the state machine. This reduces the chances of metastability in the system. See Chapter 3 for more information about metastability.

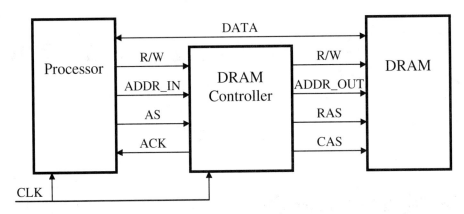

Figure 28-1 Connecting the processor to the DRAM.

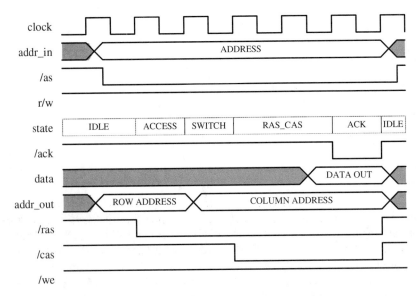

Figure 28-2 DRAM read access.

The basic DRAM accesses are shown in Figures 28-2, 28-3, and 28-4. This state machine implements early write access, which means that the write enable signal is asserted at the beginning of the access. The write takes place when the CAS signal is asserted. This is the easiest and most efficient implementation for a DRAM controller. All DRAMs need to be refreshed periodically. This controller implements a CAS-before-RAS refresh, which is the most convenient to perform because it does not require use of the address bus. Other refresh methods are better suited to non-generic controllers which may be able to generate addresses on the address bus or which can guarantee a minimum time between accesses of the DRAM.

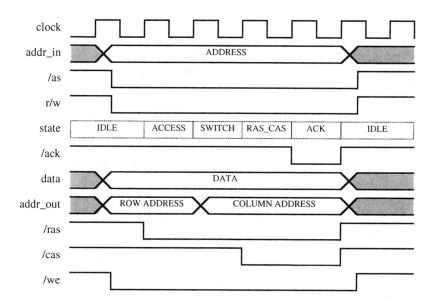

Figure 28-3 DRAM early write access.

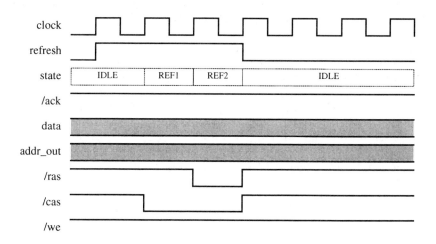

Figure 28-4 DRAM CAS-before-RAS refresh.

The state machine for the controller is shown in Figure 28-5. From the IDLE state, if there is no refresh needed and there is a DRAM access request (AS asserted by the processor), the machine enters the ACCESS state. On the next clock cycle, the controller enters the SWITCH state, which is needed to switch the address input to the DRAM from the row address to the column address. This state provides the necessary hold time for the row address to the RAS signal and enough setup time from the column address to the CAS signal. It is assumed that one clock cycle is enough. For the transition into the SWITCH state, the counter is loaded with a value representing the number of wait cycles needed for before asserting CAS. The exact number of cycles depends on the controller clock frequency and the requirements of the particular DRAM that you are using. The controller stays in this state, counting down each cycle, until the counter reaches zero. At that time, it enters the RAS_CAS state, during which RAS and CAS are both asserted. During this transition, another value is loaded into the counter. When the counter again reaches zero, it enters the ACK state during which it gives an acknowledge to the processor that the access has completed. It then returns to the IDLE state. The number of cycles needed to assert RAS and CAS, in order to meet the minimum timing requirements of the DRAM, are controlled with `define` statements in the code.

To implement the refresh function, a counter in the controller is loaded with a value representing the number of cycles to wait before performing a refresh. The exact number of cycles again depends on the controller clock frequency and the requirements of the particular DRAM that you are using. When a refresh is to be performed, it has priority over any access request by the processor. The controller enters the REF1 state and asserts the CAS signal. It then waits the appropriate number of clock cycles before entering the REF2 state and asserting the RAS signal. Again it waits the appropriate number of clock cycles before returning to the IDLE state.

Note that all of the states can be defined using the RAS, CAS, and ACK signals plus one state variable to differentiate between the RAS_CAS and REF2 states, and the ACCESS and SWITCH states. This simplifies the controller design and reduces the amount of logic needed to implement the controller.

One scheme that may increase the efficiency of the DRAM controller is to stack up refresh requests. This is done by using a counter to count the number of refresh requests. Then, DRAM accesses take priority over refresh requests, until the number of pending refresh requests reaches a critical threshold. At that time, the current DRAM access is terminated normally and all pending refresh requests are performed. This can be done as long as the maximum time between refreshes specified by the manufacturer is not violated. This type of scheme is useful when the processor (or any other device) accesses the DRAM in long bursts. The time between the bursts is used for refreshing the DRAM, without any performance penalty. This scheme is left as an "exercise for the reader" (in other words, it's getting to the end of the book, I'm burning out, and I figured it's time for you to do some of the work).

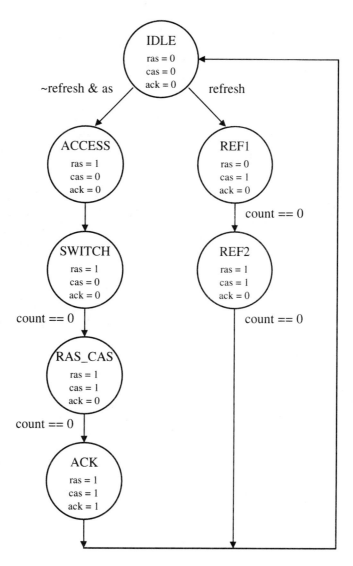

Figure 28-5 DRAM controller state machine.

28.1 BEHAVIORAL CODE

```
/*************************************************************/
// MODULE:         DRAM Controller
//
// FILE NAME:      dramcon_beh.v
// VERSION:        1.1
// DATE:           January 1, 2003
// AUTHOR:         Bob Zeidman, Zeidman Consulting
//
// CODE TYPE:      Behavioral Level
//
// DESCRIPTION:  This module implements a controller for a
// DRAM. It performs CAS-before-RAS refreshes.
//
/*************************************************************/

// DEFINES
`define DEL   1         // Clock-to-output delay. Zero
                        // time delays can be confusing
                        // and sometimes cause problems.
`define RBC_CYC 2       // Number of cycles to assert RAS
                        // before asserting CAS
`define CBR_CYC 1       // Number of cycles to assert CAS
                        // before asserting RAS
`define RACW_CYC 1      // Number of cycles to assert RAS
                        // and CAS together for a write
`define RACR_CYC 2      // Number of cycles to assert RAS
                        // and CAS together for a read
`define RACRF_CYC 1     // Number of cycles to assert RAS
                        // and CAS together for a refresh
`define CNT_BITS 2      // Number of bits needed for the
                        // counter to count the cycles
                        // listed above
`define REF_CNT 24      // Number of cycles between refreshes
`define REF_BITS 5      // Number of bits needed for the
                        // counter to count the cycles
                        // for a refresh
`define AOUT 4          // Address bit width to DRAM
`define AIN 2*`AOUT     // Address bit width from processor

// TOP MODULE
module dram_control(
      clock,
      reset_n,
```

```
        as_n,
        addr_in,
        addr_out,
        rw,
        we_n,
        ras_n,
        cas_n,
        ack);

// PARAMETERS
// These bits represent the following signals
//       col_out,ras,cas,ack
parameter[3:0]          // State machine states
    IDLE        = 4'b0000,
    ACCESS      = 4'b0100,
    SWITCH      = 4'b1100,
    RAS_CAS     = 4'b1110,
    ACK         = 4'b1111,
    REF1        = 4'b0010,
    REF2        = 4'b0110;

// INPUTS
input            clock;        // State machine clock
input            reset_n;      // Active low, synchronous reset
input            as_n;         // Active low address strobe
input [`AIN-1:0] addr_in;      // Address from processor
input            rw;           // Read/write input
                               // = 1 to read
                               // = 0 to write

// OUTPUTS
output [`AOUT-1:0]  addr_out;  // Address to DRAM
output           we_n;         // Write enable output
output           ras_n;        // Row Address Strobe to memory
output           cas_n;        // Column Address Strobe
                               // to memory
output           ack;          // Acknowledge signal
                               // to processor

// INOUTS

// SIGNAL DECLARATIONS
wire             clock;
wire             reset_n;
wire [`AIN-1:0]  addr_in;
wire             as_n;
wire             rw;
wire             we_n;
```

```
wire                 ras_n;
wire                 cas_n;
wire                 ack;
wire [`AOUT-1:0]     addr_out;

reg  [3:0]              mem_state;      // State machine
wire                 col_out;     // Output column address
                                  // = 1 for column address
                                  // = 0 for row address
reg [`CNT_BITS-1:0] count;        // Cycle counter
reg [`REF_BITS-1:0] ref_count;     // Refresh counter
reg                 refresh;      // Refresh request
// ASSIGN STATEMENTS
// Create the outputs from the states
assign col_out = mem_state[3];
assign ras_n = ~mem_state[2];
assign cas_n = ~mem_state[1];
assign ack = mem_state[0];

// Deassert we_n high during refresh
assign #`DEL we_n = rw | (mem_state == REF1) |
                (mem_state == REF2);

// Give the row address or column address to the DRAM
assign #`DEL addr_out = col_out ? addr_in[`AOUT-1:0] :
                addr_in[`AIN-1:`AOUT];

// MAIN CODE

// Look at the edge of reset
always @(reset_n) begin
    if (~reset_n) begin
        #`DEL assign mem_state = IDLE;
        assign count = `CNT_BITS'h0;
        assign ref_count = `REF_CNT;
        assign refresh = 1'b0;
    end
    else begin
        #`DEL;
        deassign mem_state;
        deassign count;
        deassign ref_count;
        deassign refresh;
    end
end
```

```verilog
// Look at the rising edge of clock for state transitions
always @(posedge clock) begin
    // Time for a refresh request?
    if (ref_count == 0) begin
        refresh <= #`DEL 1'b1;
        ref_count <= #`DEL `REF_CNT;
    end
    else
        ref_count <= #`DEL ref_count - 1;

    // Decrement cycle counter to zero
    if (count)
        count <= #`DEL count - 1;

    case (mem_state)
        IDLE:  begin
            // Refresh request has highest priority
            if (refresh) begin
                // Load the counter to assert CAS
                count <= #`DEL `CBR_CYC;
                mem_state <= #`DEL REF1;
            end
            else if (~as_n) begin
                // Load the counter to assert RAS
                count <= #`DEL `RBC_CYC;
                mem_state <= #`DEL ACCESS;
            end
        end
        ACCESS:  begin
            mem_state <= #`DEL SWITCH;
        end
        SWITCH:  begin
            if (count == 0) begin
                mem_state <= #`DEL RAS_CAS;
                if (rw)
                    count <= #`DEL `RACR_CYC;
                else
                    count <= #`DEL `RACW_CYC;
            end
        end
        RAS_CAS:begin
            if (count == 0) begin
                mem_state <= #`DEL ACK;
            end
        end
        ACK:   begin
            mem_state <= #`DEL IDLE;
        end
```

```
      REF1:  begin
         if (count == 0) begin
            mem_state <= #`DEL REF2;
            count <= #`DEL `RACRF_CYC;
         end
      end
      REF2:  begin
         if (count == 0) begin
            mem_state <= #`DEL IDLE;
            refresh <= #`DEL 1'b0;
         end
      end
   endcase
end
endmodule     // dram_control
```

28.2 RTL CODE

```
/**********************************************************/
// MODULE:       DRAM Controller
//
// FILE NAME:    dramcon_rtl.v
// VERSION:      1.1
// DATE:         January 1, 2003
// AUTHOR:       Bob Zeidman, Zeidman Consulting
//
// CODE TYPE:    Register Transfer Level
//
// DESCRIPTION:  This module implements a controller for a
// DRAM. It performs CAS-before-RAS refreshes.
//
/**********************************************************/

// DEFINES
`define DEL   1          // Clock-to-output delay. Zero
                         // time delays can be confusing
                         // and sometimes cause problems.
`define RBC_CYC 2        // Number of cycles to assert RAS
                         // before asserting CAS
`define CBR_CYC 1        // Number of cycles to assert CAS
                         // before asserting RAS
`define RACW_CYC 1       // Number of cycles to assert RAS
                         // and CAS together for a write
`define RACR_CYC 2       // Number of cycles to assert RAS
```

```
                        // and CAS together for a read
`define RACRF_CYC 1     // Number of cycles to assert RAS
                        // and CAS together for a refresh
`define CNT_BITS 2      // Number of bits needed for the
                        // counter to count the cycles
                        // listed above
`define REF_CNT 24      // Number of cycles between refreshes
`define REF_BITS 5      // Number of bits needed for the
                        // counter to count the cycles
                        // for a refresh
`define AOUT 4          // Address bit width to DRAM
`define AIN 2*`AOUT     // Address bit width from processor

// TOP MODULE
module dram_control(
      clock,
      reset_n,
      as_n,
      addr_in,
      addr_out,
      rw,
      we_n,
      ras_n,
      cas_n,
      ack);


// PARAMETERS
// These bits represent the following signals
//        col_out,ras,cas,ack
parameter[3:0]          // State machine states
   IDLE      = 4'b0000,
   ACCESS    = 4'b0100,
   SWITCH    = 4'b1100,
   RAS_CAS   = 4'b1110,
   ACK       = 4'b1111,
   REF1      = 4'b0010,
   REF2      = 4'b0110;

// INPUTS
input                clock;    // State machine clock
input                reset_n;  // Active low, synchronous reset
input                as_n;     // Active low address strobe
input [`AIN-1:0]     addr_in;  // Address from processor
input                rw;       // Read/write input
                               // = 1 to read
                               // = 0 to write
```

```verilog
// OUTPUTS
output [`AOUT-1:0]  addr_out;  // Address to DRAM
output              we_n;      // Write enable output
output              ras_n;     // Row Address Strobe to memory
output              cas_n;     // Column Address Strobe
                               // to memory
output              ack;       // Acknowledge signal
                               // to processor


// INOUTS


// SIGNAL DECLARATIONS
wire                clock;
wire                reset_n;
wire [`AIN-1:0]     addr_in;
wire                as_n;
wire                rw;
wire                we_n;
wire                ras_n;
wire                cas_n;
wire                ack;
wire [`AOUT-1:0]    addr_out;

reg [3:0]           mem_state;      // Synthesis state_machine
wire                col_out;   // Output column address
                               // = 1 for column address
                               // = 0 for row address
reg [`CNT_BITS-1:0] count;     // Cycle counter
reg [`REF_BITS-1:0] ref_count; // Refresh counter
reg                 refresh;   // Refresh request

// ASSIGN STATEMENTS
// Create the outputs from the states
assign col_out = mem_state[3];
assign ras_n = ~mem_state[2];
assign cas_n = ~mem_state[1];
assign ack = mem_state[0];

// Deassert we_n high during refresh
assign #`DEL we_n = rw | (mem_state == REF1) |
                (mem_state == REF2);

// Give the row address or column address to the DRAM
assign #`DEL addr_out = col_out ? addr_in[`AOUT-1:0] :
                addr_in[`AIN-1:`AOUT];
```

```
// MAIN CODE

// Look at the rising edge of clock for state transitions
always @(posedge clock or negedge reset_n) begin
    if (~reset_n) begin        ·
        mem_state <= #`DEL IDLE;
        count <= #`DEL `CNT_BITS'h0;
        ref_count <= #`DEL `REF_CNT;
        refresh <= #`DEL 1'b0;
    end
    else begin
        // Time for a refresh request?
        if (ref_count == 0) begin
            refresh <= #`DEL 1'b1;
            ref_count <= #`DEL `REF_CNT;
        end
        else
            ref_count <= #`DEL ref_count - 1;

        // Decrement cycle counter to zero
        if (count)
            count <= #`DEL count - 1;

        case (mem_state) // synthesis full_case parallel_case
            IDLE:  begin
                // Refresh request has highest priority
                if (refresh) begin
                    // Load the counter to assert CAS
                    count <= #`DEL `CBR_CYC;
                    mem_state <= #`DEL REF1;
                end
                else if (~as_n) begin
                    // Load the counter to assert RAS
                    count <= #`DEL `RBC_CYC;
                    mem_state <= #`DEL ACCESS;
                end
            end
            ACCESS:   begin
                mem_state <= #`DEL SWITCH;
            end
            SWITCH:   begin
                if (count == 0) begin
                    mem_state <= #`DEL RAS_CAS;
                    if (rw)
                        count <= #`DEL `RACR_CYC;
                    else
                        count <= #`DEL `RACW_CYC;
```

```
            end
        end
        RAS_CAS:begin
            if (count == 0) begin
                mem_state <= #`DEL ACK;
            end
        end
        ACK:    begin
            mem_state <= #`DEL IDLE;
        end
        REF1:   begin
            if (count == 0) begin
                mem_state <= #`DEL REF2;
                count <= #`DEL `RACRF_CYC;
            end
        end
        REF2:   begin
            if (count == 0) begin
                mem_state <= #`DEL IDLE;
                refresh <= #`DEL 1'b0;
            end
        end
        endcase
    end
end
endmodule    // dram_control
```

28.3 SIMULATION CODE

The simulation code uses a DRAM memory model. The simulation instantiates the memory controller and the DRAM. First it asserts the reset signal and checks that outputs are reset to the correct values. Then it writes a series of data words to each successive location in memory. It then reads back those locations and checks that the data words match those that were previously written. The method for testing correct timing of the signals uses code that is embedded in the DRAM code. It is the same method as that described in Chapter 26 for the SRAM model.

Note that this DRAM model is actually a Fast Page DRAM. This DRAM controller does not take advantage of the Fast Page mode. The controller in the next chapter does take advantage of this faster access mode, and it is simulated using the same simulation code as shown here.

```
/***********************************************************/
// MODULE:          DRAM Controller simulation
```

```
//
// FILE NAME:      dramcon_sim.v
// VERSION:        1.1
// DATE:           January 1, 2003
// AUTHOR:         Bob Zeidman, Zeidman Consulting
//
// CODE TYPE:      Simulation
//
// DESCRIPTION:  This module provides stimuli for simulating
// a DRAM memory controller. It uses a DRAM memory model and
// performs a number of writes and reads, including
// back-to-back reads and writes. It also checks the reset
// function.
//
/***********************************************************/

// DEFINES
`define DEL   1             // Clock-to-output delay. Zero
                            // time delays can be confusing
                            // and sometimes cause problems.
`define DEL2 2             // Longer clock-to-output delay.
`define ACC_COUNT 4        // Maximum number of cycles needed
                            // to access memory
`define CLK_HPERIOD 50     // Clock half period
                            // Clock period
`define CLK_PERIOD 2*`CLK_HPERIOD
`define WR_PERIOD 70        // Write access time of memory
                            // from assertion of CAS
`define RD_PERIOD 120       // Read access time of memory
                            // from assertion of CAS
`define AOUT 4             // Address bit width to DRAM
`define AIN 2*`AOUT        // Address bit width from processor
`define DWIDTH 8           // Data width of DRAM
`define DDEPTH 256         // Data depth of DRAM

// TOP MODULE
module dramcon_sim();

// PARAMETERS

// INPUTS

// OUTPUTS

// INOUTS
```

```
// SIGNAL DECLARATIONS
reg                     clock;
reg                     reset;

reg   [`AIN-1:0]        addr_in;
reg                     as;
reg                     rw;
wire                    we_n;
wire                    ras_n;
wire                    cas_n;
wire                    ack;
wire  [`AOUT-1:0]       addr_out;

reg [`DWIDTH-1:0]  data_out;      // Data out of processor
wire [`DWIDTH-1:0] data;          // Data bus
reg [`AIN:0]       addr_count;    // Address counter
reg [`DWIDTH-1:0]  data_patt;       // Data pattern holder
integer            cyc_count;       // Count cycles to determine
                                    // if the access has taken
                                    // too long
// ASSIGN STATEMENTS
assign data = (as & we_n) ? 8'hzz : data_out;

// MAIN CODE

// Instantiate the controller
dram_control Dram_control(
        .clock(clock),
        .reset_n(~reset),
        .as_n(~as),
        .addr_in(addr_in),
        .addr_out(addr_out),
        .rw(rw),
        .we_n(we_n),
        .ras_n(ras_n),
        .cas_n(cas_n),
        .ack(ack));
// Instantiate a DRAM
dram Dram(
        .ras(~ras_n),
        .cas(~cas_n),
        .we(~we_n),
        .address(addr_out),
        .data(data));

// Generate the clock
always #`CLK_HPERIOD clock = ~clock;
```

```verilog
// Simulate
initial begin
    // Initialize inputs
    clock = 1;
    reset = 0;
    as = 0;

    // Test the reset signal
    #`DEL;
    // Assert the reset signal
    reset = 1;

    // Wait for the outputs to change asynchronously
    #`DEL
    #`DEL
    // Test outputs
    if ((ras_n === 1'b1) && (cas_n === 1'b1) &&
        (ack === 1'b0))
        $display ("Reset is working");
    else begin
        $display("\nERROR at time %0t:", $time);
        $display("Reset is not working");
        $display("    ras_n = %b", ras_n);
        $display("    cas_n = %b", cas_n);
        $display("    ack   = %b\n", ack);

        // Use $stop for debugging
        $stop;
    end

    // Deassert the reset signal
    reset = 0;

    // Initialize the address counter
    addr_count = 0;

    // Initialize the data pattern to be written
    data_patt = `DWIDTH'h1;

    // Write a series of values to memory
    while (addr_count[`AIN] === 0) begin
        // Write to memory
        writemem(addr_count[`AIN-1:0], data_patt);

        // Increment the address counter
        addr_count <= addr_count + 1;    // Shift the data pattern
        data_patt <= (data_patt << 1);
```

```
            data_patt[0] <= data_patt[`DWIDTH-1];
            // Wait for the data pattern to change
            #`DEL;
        end

        // Initialize the address counter
        addr_count = 0;

        // Initialize the data pattern to be read
        data_patt = `DWIDTH'h1;

        // Verify the values that were written
        while (addr_count[`AIN] === 0) begin
            // Read from memory
            readmem(addr_count[`AIN-1:0], data_patt);

            // Increment the address counter
            addr_count <= addr_count + 1;

            // Shift the data pattern
            data_patt <= (data_patt << 1);
            data_patt[0] <= data_patt[`DWIDTH-1];

            // Wait for the data pattern to change
            #`DEL;
        end

        $display("\nSimulation complete - no errors\n");
        $finish;
    end

    // Check the number of cycles for each access
    always @(posedge clock) begin
        // Check whether an access is taking too long
        if (cyc_count > 3*`ACC_COUNT) begin
            $display("\nERROR at time %0t:", $time);
            if (rw)
                $display("Read access took too long\n");
            else
                $display("Write access took too long\n");

            // Use $stop for debugging
            $stop;
        end
    end
```

```
// TASKS
// Write data to memory
task writemem;

// INPUTS
input [`AIN-1:0]    write_addr;      // Memory address
input [`DWIDTH-1:0] write_data;      // Data to write to memory

// OUTPUTS

// INOUTS

// TASK CODE
begin
    cyc_count = 0;          // Initialize the cycle count

    // Wait for the rising clock edge
    @(posedge clock);
    rw <= #`DEL 0;          // Set up a write access
                            // Set up the address
    addr_in <= #`DEL write_addr;
                            // Set up the data to be written
    data_out <= #`DEL write_data;
    as <= #`DEL2 1;         // Assert address strobe

    // Wait for the acknowledge
    @(posedge clock);
    while (~ack) begin
        // Increment the cycle count
        cyc_count = cyc_count + 1;

        @(posedge clock);
    end

    as <= #`DEL 0;          // Deassert address strobe
    @(posedge clock);       // Wait one clock cycle

end
endtask        // writemem

// Read data from memory and check its value
task readmem;

// INPUTS
input [`AIN-1:0]    read_addr;    // Memory address
input [`DWIDTH-1:0] exp_data;     // Expected read data
```

```verilog
// OUTPUTS

// INOUTS

// TASK CODE
begin
    cyc_count = 0;            // Initialize the cycle count

    // Wait for the rising clock edge
    @(posedge clock);
    rw <= #`DEL 1;            // Set up a read access
                             // Set up the address
    addr_in <= #`DEL read_addr;
    as <= #`DEL2 1;          // Assert address strobe

    // Wait for the acknowledge
    @(posedge clock);
    while (~ack) begin
       // Increment the cycle count
       cyc_count = cyc_count + 1;

       @(posedge clock);
    end

    // Did we find the expected data?
    if (data !== exp_data) begin
       $display("\nERROR at time %0t:", $time);
       $display("Controller is not working");
       $display("    data written = %h", exp_data);
       $display("    data read    = %h\n", data);

       // Use $stop for debugging
       $stop;
    end

    as <= #`DEL 0;           // Deassert address strobe
    @(posedge clock);        // Wait one clock cycle

end
endtask         // readmem

endmodule          // dramcon_sim

// SUBMODULE
// DRAM memory model
module dram(
      ras,
```

```
        cas,
        we,
        address,
        data);

// INPUTS
input                   ras;        // Row address strobe
input                   cas;        // Column address strobe
input                   we;         // Write enable input
input [`AOUT-1:0]       address;    // Address input to DRAM

// OUTPUTS

// INOUTS
inout [`DWIDTH-1:0] data;           // Data lines

// SIGNAL DECLARATIONS
wire                    ras;
wire                    cas;
wire                    we;
wire [`DWIDTH-1:0]      data;
reg  [`DWIDTH-1:0]      mem[`DDEPTH-1:0];   // Stored data
reg  [`AIN-1:0]         mem_addr;           // Memory address

time                    rd_time;    // Time of the most recent change of
                                    // the rd input
wire [63:0]             rd_dtime;   // rd_time delayed by `RD_PERIOD
                                    // time units
time                    wr_time;    // Time of the most recent change of
                                    // the wr input
wire [63:0]             wr_dtime;   // wr_time delayed by `WD_PERIOD
                                    // time units

reg                     oen;        // Internal output enable - this
                                    // signal is asserted after the
                                    // minimum CAS time for a read
reg                     wen;        // Internal write enable - this
                                    // signal is asserted after the
                                    // minimum CAS time for a write

// PARAMETERS

// ASSIGN STATEMENTS
assign #`RD_PERIOD rd_dtime = rd_time;
assign #`WR_PERIOD wr_dtime = wr_time;

// Output the data if the we signal is high and the oe is
```

```
// asserted, which means that we met the minimum CAS time
// for a read
assign data = (oen & ~we) ? mem[address] : 8'hzz;

// MAIN CODE
initial begin
   oen = 0;
   wen = 0;
   wr_time = 0;
   rd_time = 0;
end

// Look for the RAS rising edge for the row address
always @(posedge ras) begin
   mem_addr[`AIN-1:`AOUT] <= #`DEL address;
end

// Look for the CAS rising edge for the column address
always @(posedge cas) begin
   mem_addr[`AOUT-1:0] <= #`DEL address;
end

// Look for the beginning of the access
always @(posedge cas) begin
   if (we) begin
                            // This is a write
      wr_time = $time;   // Save the time of the most
                         // recent rising edge of wr
      wen = 0;           // Deassert internal write enable
   end
   else begin
                            // This is a read
      rd_time = $time;   // Save the time of the most
                         // recent rising edge of rd
      oen = 0;           // Deassert internal read enable
   end
end

// Determine whether to assert the internal write enable
always @(wr_dtime) begin
   if (wr_time === wr_dtime) begin
      // If this is true, the last rising edge of CAS
      // was exactly `WR_PERIOD time units ago, so
      // assert the internal write enable
      wen = 1;
   end
end
```

```verilog
// Determine whether to assert the internal read enable
always @(rd_dtime) begin
    if (rd_time === rd_dtime) begin
        // If this is true, the last rising edge of rd
        // was exactly `RD_PERIOD time units ago, so
        // assert the internal output enable
        oen = 1;
    end
end

// Look for the end of the access
always @(negedge cas) begin
    if (wen & we) begin
        // Write the data if the write enable is asserted
        // which means we met the minimum wr pulse time
        mem[mem_addr] <= #`DEL data;
    end
    wen = 0;            // Turn off the internal write enable
    oen = 0;            // Turn off the internal output enable
end

endmodule    // dram
```

The Fast Page Mode DRAM Controller

*T*his chapter describes a controller for a Fast Page Mode Dynamic Random Access Memory. The difference between a fast page DRAM and a normal DRAM is that the fast page DRAM keeps the entire page of data, addressed by the row address, active as long as the Row Address Strobe (RAS) is kept asserted. When a new access is required to that same page, only the CAS address needs to be supplied while toggling the Column Address Strobe (CAS). These page accesses are very fast, as the name implies, with respect to a normal DRAM access. Typically they require about half the normal access time. The concept here is that most processor accesses occur to data that is localized. In other words, if a particular memory location is accessed at one time, the next access has a very high chance of being very close to the first. This is true for data that consists of strings, vectors, arrays, or video frames. For instructions, most instructions are sequential except in the case of occasional branches or interrupts. In most systems, the use of Fast Page Mode DRAMs greatly increases overall system performance.

The connection between the processor, the DRAM, and the controller looks exactly like that for an ordinary DRAM, as shown in the previous chapter. The Row Address Strobe

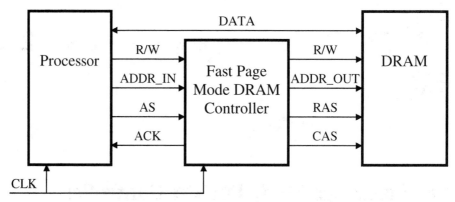

Figure 29-1 Connecting the Processor to the Fast Page Mode DRAM

(RAS) and Column Address Strobe (CAS) to the DRAM, and the Acknowledge input to the processor (ACK) are derived from the processor Address Strobe (AS) by the controller. Also, the controller must include a periodic refresh of the DRAM using the RAS and CAS inputs.

As before, I assume that the signals from the processor are synchronized to the clock signal. If this is not the case, then the address strobe should be synchronized to the clock by sending it through at least one flip-flop, before it is used in the state machine. This reduces the chances of metastability in the system. See Chapter 3 for more information about metastability.

The fast page DRAM accesses are shown in Figures 29-2, 29-3, and 29-4. As with the normal DRAM controller, this state machine implements early write access. This controller also implements a CAS-before-RAS refresh.

The state machine for the controller is shown in Figure 29-5. From the IDLE state, if there is no refresh needed and there is a DRAM access request (AS asserted by the processor), the machine enters the ACCESS state. On the next clock cycle, the controller enters the SWITCH state, which is needed to switch the address input to the DRAM from the row address to the column address. This state provides the necessary hold time for the row address to the RAS signal and setup time from the column address to the CAS signal. It is assumed that one clock cycle is enough. For the transition into the SWITCH state, the counter is loaded with a value representing the number of wait cycles needed for before asserting CAS. The exact number of cycles depends on the controller clock frequency and the requirements of the particular DRAM that you are using. The controller stays in this state, counting down each cycle, until the counter reaches zero. At that time, it enters the RAS_CAS state, during which RAS and CAS are both asserted. During this transition, another value is loaded into the counter. When the counter again reaches zero, it enters the ACK state during which it gives an acknowledge to the processor that the access has completed. If a refresh request is pending, the controller goes to the IDLE state in order to perform the refresh. Otherwise, rather than returning to the IDLE state, as it would for a normal DRAM, the controller enters the WAIT state and keeps the RAS signal asserted. If a refresh is requested, the controller returns to the IDLE state to perform the refresh. If the address strobe is asserted, indicating that the processor is attempting another DRAM access, the controller compares the previous row address to the current row address. If they do not match, the controller returns to the IDLE state to perform a completely new access. If, however, they do match, then this current access is on the same page as the previous access. In that case, the controller goes to the RAS_CAS state to toggle the CAS signal for this new fast page access.

Note that the refresh function and the refresh cycle timing for this controller is identical to that of the normal DRAM controller discussed in the previous chapter.

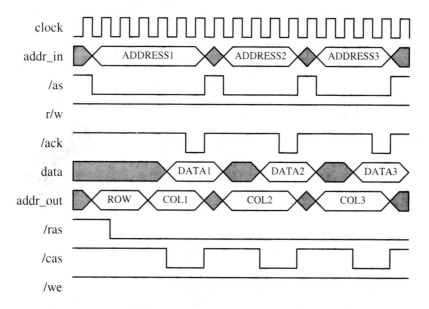

Figure 29-2 Fast page mode DRAM read access.

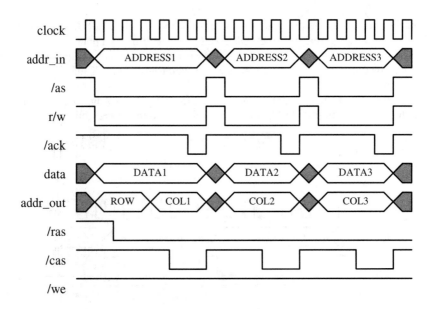

Figure 29-3 Fast page mode DRAM early write access.

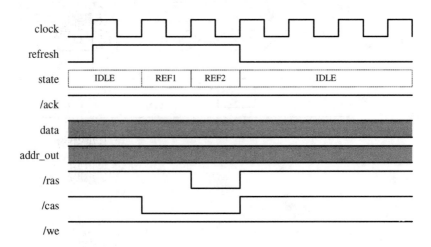

Figure 29-4 DRAM CAS-before-RAS refresh.

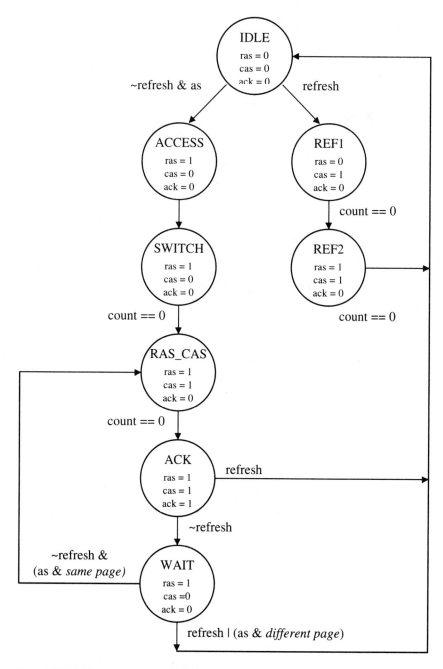

Figure 29-5 Fast page mode DRAM controller state machine.

29.1 BEHAVIORAL CODE

```
/**********************************************************/
// MODULE:          Fast Page Mode DRAM Controller
//
// FILE NAME:       fpdramcon_beh.v
// VERSION:         1.1
// DATE:            January 1, 2003
// AUTHOR:          Bob Zeidman, Zeidman Consulting
//
// CODE TYPE:       Behavioral Level
//
// DESCRIPTION:  This module implements a controller for a
// Fast Page Mode DRAM. It performs CAS-before-RAS refreshes.
//
/**********************************************************/

// DEFINES
`define DEL   1           // Clock-to-output delay. Zero
                          // time delays can be confusing
                          // and sometimes cause problems.
`define RBC_CYC 2         // Number of cycles to assert RAS
                          // before asserting CAS
`define CBR_CYC 1         // Number of cycles to assert CAS
                          // before asserting RAS
`define RACW_CYC 1        // Number of cycles to assert RAS
                          // and CAS together for a write
`define RACR_CYC 2        // Number of cycles to assert RAS
                          // and CAS together for a read
`define RACRF_CYC 1       // Number of cycles to assert RAS
                          // and CAS together for a refresh
`define CNT_BITS 2        // Number of bits needed for the
                          // counter to count the cycles
                          // listed above
`define REF_CNT 24        // Number of cycles between refreshes
`define REF_BITS 5        // Number of bits needed for the
                          // counter to count the cycles
                          // for a refresh
`define AOUT 4            // Address bit width to DRAM
`define AIN 2*`AOUT       // Address bit width from processor

// TOP MODULE
module dram_control(
     clock,
     reset_n,
```

```
        as_n,
        addr_in,
        addr_out,
        rw,
        we_n,
        ras_n,
        cas_n,
        ack);

// PARAMETERS
// These bits represent the following signals
//        state_bit,col_out,ras,cas,ack
parameter[4:0]           // State machine states
    IDLE      = 5'b00000,
    ACCESS    = 5'b00100,
    SWITCH    = 5'b01100,
    RAS_CAS   = 5'b01110,
    ACK       = 5'b01111,
    WAIT      = 5'b11100,
    REF1      = 5'b00010,
    REF2      = 5'b00110;

// INPUTS
input                   clock;    // State machine clock
input                   reset_n;  // Active low, synchronous reset
input                   as_n;     // Active low address strobe
input [`AIN-1:0]        addr_in;  // Address from processor
input                   rw;       // Read/write input
                                  // = 1 to read
                                  // = 0 to write

// OUTPUTS
output [`AOUT-1:0]      addr_out; // Address to DRAM
output                  we_n;     // Write enable output
output                  ras_n;    // Row Address Strobe to memory
output                  cas_n;    // Column Address Strobe
                                  // to memory
output                  ack;      // Acknowledge signal
                                  // to processor

// INOUTS

// SIGNAL DECLARATIONS
wire                    clock;
wire                    reset_n;
wire [`AIN-1:0]         addr_in;
wire                    as_n;
wire                    rw;
```

```
wire                  we_n;
wire                  ras_n;
wire                  cas_n;
wire                  ack;
wire [`AOUT-1:0]      addr_out;

reg   [4:0]              mem_state;        // State machine
wire                    col_out;      // Output column address
                                      // = 1 for column address
                                      // = 0 for row address
reg  [`CNT_BITS-1:0]    count;            // Cycle counter
reg  [`REF_BITS-1:0]    ref_count;        // Refresh counter

reg                     refresh;      // Refresh request
reg  [`AOUT-1:0]    last_addr;        // Most significant bits of
                                      // address of last access

// ASSIGN STATEMENTS
// Create the outputs from the states
assign col_out = mem_state[3];
assign ras_n = ~mem_state[2];
assign cas_n = ~mem_state[1];
assign ack = mem_state[0];

// Deassert we_n high during refresh
assign #`DEL we_n = rw | (mem_state == REF1) |
               (mem_state == REF2);

// Give the row address or column address to the DRAM
assign #`DEL addr_out = col_out ? addr_in[`AOUT-1:0] :
                        addr_in[`AIN-1:`AOUT];

// MAIN CODE

// Look at the edge of reset
always @(reset_n) begin
   if (~reset_n) begin
       #`DEL assign mem_state = IDLE;
       assign count = `CNT_BITS'h0;
       assign ref_count = `REF_CNT;
       assign refresh = 1'b0;
   end
   else begin
       #`DEL;
       deassign mem_state;
       deassign count;
       deassign ref_count;
```

```
            deassign refresh;
        end
    end
end

// Look at the rising edge of clock for state transitions
always @(posedge clock) begin
    // Time for a refresh request?
    if (ref_count == 0) begin
        refresh <= #`DEL 1'b1;
        ref_count <= #`DEL `REF_CNT;
    end
    else
        ref_count <= #`DEL ref_count - 1;

    // Decrement cycle counter to zero
    if (count)
        count <= #`DEL count - 1;

        case (mem_state)
        IDLE:  begin
            // Refresh request has highest priority
            if (refresh) begin
                // Load the counter to assert CAS
                count <= #`DEL `CBR_CYC;
                mem_state <= #`DEL REF1;
            end
            else if (~as_n) begin
                // Load the counter to assert RAS
                count <= #`DEL `RBC_CYC;
                mem_state <= #`DEL ACCESS;
                // Save the address bits
                last_addr <= #`DEL addr_in[`AIN-1:`AOUT];
            end
        end
        ACCESS:   begin
            mem_state <= #`DEL SWITCH;
        end
        SWITCH:   begin
            if (count == 0) begin
                mem_state <= #`DEL RAS_CAS;
                if (rw)
                    count <= #`DEL `RACR_CYC;
                else
                    count <= #`DEL `RACW_CYC;
            end
        end
        RAS_CAS:begin
            if (count == 0) begin
```

```
                       mem_state <= #`DEL ACK;
                  end
              end
              ACK:  begin
                  if (refresh)
                      mem_state <= #`DEL IDLE;
                  else
                      mem_state <= #`DEL WAIT;
              end
              WAIT:  begin
                  if (refresh)
                      mem_state <= #`DEL IDLE;
                  else if (~as_n) begin
                      // Is this the same page as
                      // the previous access?
                      if (addr_in[`AIN-1:`AOUT] == last_addr)
                          mem_state <= #`DEL RAS_CAS;
                      else
                          mem_state <= #`DEL IDLE;
                  end
              end
              REF1:  begin
                  if (count == 0) begin
                      mem_state <= #`DEL REF2;
                      count <= #`DEL `RACRF_CYC;
                  end
              end
              REF2:  begin
                  if (count == 0) begin
                      mem_state <= #`DEL IDLE;
                      refresh <= #`DEL 1'b0;
                  end
              end
          endcase
      end
      endmodule     // dram_control
```

29.2 RTL Code

```
/***********************************************************/
// MODULE:          Fast Page Mode DRAM Controller
//
// FILE NAME:       fpdramcon_rtl.v
// VERSION:         1.1
```

```
// DATE:        January 1, 2003
// AUTHOR:      Bob Zeidman, Zeidman Consulting
//
// CODE TYPE:   Register Transfer Level
//
// DESCRIPTION: This module implements a controller for a
// Fast Page Mode DRAM. It performs CAS-before-RAS refreshes.
//
/*********************************************************/

// DEFINES
`define DEL   1          // Clock-to-output delay. Zero
                         // time delays can be confusing
                         // and sometimes cause problems.
`define RBC_CYC 2        // Number of cycles to assert RAS
                         // before asserting CAS
`define CBR_CYC 1        // Number of cycles to assert CAS
                         // before asserting RAS
`define RACW_CYC 1       // Number of cycles to assert RAS
                         // and CAS together for a write
`define RACR_CYC 2       // Number of cycles to assert RAS
                         // and CAS together for a read
`define RACRF_CYC 1      // Number of cycles to assert RAS
                         // and CAS together for a refresh
`define CNT_BITS 2       // Number of bits needed for the
                         // counter to count the cycles
                         // listed above
`define REF_CNT 24       // Number of cycles between refreshes
`define REF_BITS 5       // Number of bits needed for the
                         // counter to count the cycles
                         // for a refresh
`define AOUT 4           // Address bit width to DRAM
`define AIN 2*`AOUT      // Address bit width from processor

// TOP MODULE
module dram_control(
     clock,
     reset_n,
     as_n,
     addr_in,
     addr_out,
     rw,
     we_n,
     ras_n,
     cas_n,
     ack);
```

```
// PARAMETERS
// These bits represent the following signals
//         state_bit,col_out,ras,cas,ack
parameter[4:0]              // State machine states
    IDLE      = 5'b00000,

    ACCESS    = 5'b00100,
    SWITCH    = 5'b01100,
    RAS_CAS   = 5'b01110,
    ACK       = 5'b01111,
    WAIT      = 5'b11100,
    REF1      = 5'b00010,
    REF2      = 5'b00110;

// INPUTS
input                clock;      // State machine clock
input                reset_n;    // Active low, synchronous reset
input                as_n;       // Active low address strobe
input [`AIN-1:0]     addr_in;    // Address from processor
input                rw;         // Read/write input
                                 // = 1 to read
                                 // = 0 to write

// OUTPUTS
output [`AOUT-1:0]   addr_out;   // Address to DRAM
output               we_n;       // Write enable output
output               ras_n;      // Row Address Strobe to memory
output               cas_n;      // Column Address Strobe
                                 // to memory
output               ack;        // Acknowledge signal
                                 // to processor

// INOUTS

// SIGNAL DECLARATIONS
wire                 clock;
wire                 reset_n;
wire [`AIN-1:0]      addr_in;
wire                 as_n;
wire                 rw;
wire                 we_n;
wire                 ras_n;
wire                 cas_n;
wire                 ack;
wire [`AOUT-1:0]     addr_out;

reg  [4:0]           mem_state;      // Synthesis state_machine
wire                 col_out;    // Output column address
```

```
                               // = 1 for column address
                               // = 0 for row address
reg  [`CNT_BITS-1:0]    count;         // Cycle counter
reg  [`REF_BITS-1:0]    ref_count;        // Refresh counter
reg                     refresh;       // Refresh request
reg  [`AOUT-1:0]    last_addr;        // Most significant bits of
                               // address of last access

// ASSIGN STATEMENTS
// Create the outputs from the states
assign col_out = mem_state[3];
assign ras_n = ~mem_state[2];
assign cas_n = ~mem_state[1];
assign ack = mem_state[0];

// Deassert we_n high during refresh
assign #`DEL we_n = rw | (mem_state == REF1) |
                (mem_state == REF2);

// Give the row address or column address to the DRAM
assign #`DEL addr_out = col_out ? addr_in[`AOUT-1:0] :
                addr_in[`AIN-1:`AOUT];

// MAIN CODE

// Look at the rising edge of clock for state transitions
always @(posedge clock or negedge reset_n) begin
   if (~reset_n) begin
      mem_state <= #`DEL IDLE;
      count <= #`DEL `CNT_BITS'h0;
      ref_count <= #`DEL `REF_CNT;
      refresh <= #`DEL 1'b0;
      last_addr <= #`DEL `AOUT'h0;
   end
   else begin
      // Time for a refresh request?
      if (ref_count == 0) begin
         refresh <= #`DEL 1'b1;
         ref_count <= #`DEL `REF_CNT;
      end
      else
         ref_count <= #`DEL ref_count - 1;

      // Decrement cycle counter to zero
      if (count)
         count <= #`DEL count - 1;
```

```
case (mem_state)
   IDLE:  begin
      // Refresh request has highest priority
      if (refresh) begin
         // Load the counter to assert CAS
         count <= #`DEL `CBR_CYC;
         mem_state <= #`DEL REF1;
      end
      else if (~as_n) begin
         // Load the counter to assert RAS
         count <= #`DEL `RBC_CYC;
         mem_state <= #`DEL ACCESS;

         // Save the address bits
         last_addr <= #`DEL addr_in[`AIN-1:`AOUT];
      end
   end
   ACCESS:   begin
      mem_state <= #`DEL SWITCH;
   end
   SWITCH:   begin
      if (count == 0) begin
         mem_state <= #`DEL RAS_CAS;
         if (rw)
            count <= #`DEL `RACR_CYC;
         else
            count <= #`DEL `RACW_CYC;
      end
   end
   RAS_CAS:begin
      if (count == 0) begin
         mem_state <= #`DEL ACK;
      end
   end
   ACK:   begin
      if (refresh)
         mem_state <= #`DEL IDLE;
      else
         mem_state <= #`DEL WAIT;
   end
   WAIT:  begin
      if (refresh)
         mem_state <= #`DEL IDLE;
      else if (~as_n) begin
         // Is this the same page as
         // the previous access?
         if (addr_in[`AIN-1:`AOUT] == last_addr)
```

```
                          mem_state <= #`DEL RAS_CAS;
                  else
                          mem_state <= #`DEL IDLE;
              end
          end
          REF1:  begin
              if (count == 0) begin
                  mem_state <= #`DEL REF2;
                  count <= #`DEL `RACRF_CYC;
              end
          end
          REF2:  begin
              if (count == 0) begin
                  mem_state <= #`DEL IDLE;
                  refresh <= #`DEL 1'b0;
              end
          end
      endcase
   end
end
endmodule    // dram_control
```

29.3 SIMULATION CODE

The simulation code for the Fast Page DRAM is exactly the same as the simulation code for the DRAM in the previous chapter. This is because the DRAM model used for that simulation is a Fast Page DRAM.

Appendix A

RESOURCES

*T*his appendix contains books, magazines and web sites that I used as references while writing this text. It also contains other references that I feel will be useful to the reader.

GENERAL ELECTRICAL ENGINEERING

Dorf, Richard C., *The Electrical Engineering Handbook*. Boca Raton, FL: CRC Press, 1993.

Keating, Michael and Bricaud, Pierre, *Reuse Methodology Manual for System-On-A-Chip Designs*. Norwell, MA: Kluwer Academic Publishers, 1998.

Kuo, Franklin F., *An Introduction to Computer Logic*. Englewood Cliffs, NJ: Prentice-Hall, Inc., 1975.

Maxfield, Clive, *Designus Maximus Unleashed!* Boston, MA: Newnes, 1998.

Stone, Harold S., *Introduction to Computer Architecture*, Second Edition. Chicago, IL: Science Research Associates, Inc., 1980.

Texas Instruments, *TTL Logic Data Book. Dallas*, TX: Texas Instruments, Inc., 1988.

INTEGRATED CIRCUITS

Mead, Carver and Conway, Lynn, *Introduction to VLSI Systems*. Reading, MA: Addison-Wesley Publishing Co., 1980.

Zeidman, Bob, *Designing with FPGAs and CPLDs*, Lawrence, KS: CMP Books, 2002

VERILOG

IEEE Computer Society, *IEEE Std 1364-1995: IEEE Standard Hardware Description Language Based on the Verilog® Hardware Description Language*. New York, NY: Institute of Electrical and Electronic Engineers, 1996.

Sternheim, Eliezer, et al., *Digital Design with Verilog® HDL*. Cupertino, CA: Automata Publishing Company, 1990.

Glossary

ASIC Application Integrated Circuit. A chip that is designed by a customer for proprietary use. It is designed using off-the-shelf tools, but is produced by a specific semiconductor vendor using their manufacturing process.

Behavioral Level A method of coding hardware that uses high-level representations of algorithms and general functionality without considering low-level issues like clocking, registers, and gate delays.

BIST Built-In Self Test. Circuitry on a chip that is used to test the functionality of the chip.

CPLD Complex Programmable Logic Device. A programmable device that has an architecture similar to a large number of PALs.

CRC Cyclic Redundancy Check. A method of detecting errors in long sequences of data, using an FCS.

EDC Error Correction and Detection. Codes that can detect a certain number of single-bit errors in a sequence of data bits, and correct a number of those errors.

FCS Frame Check Sequence. The sequence of bits appended to a sequence of data bits for error detection using a CRC.

FPGA Field Programmable Gate Array. A programmable device that has an architecture similar to a gate array ASIC.

FSM Finite State Machine. A state machine with a finite number of states.

Gate Array A specific type of ASIC architecture that has a large number of transistors arranged in an array. Metal layers are added by the designers to create specific custom functions.

Hamming Codes EDC codes that rely on the work of R.W. Hamming.

HDL   Hardware Description Language. A high-level, programming-like language used for designing hardware.

LFSR   Linear Feedback Shift Register. A shift register that shifts the bits through specific XOR gates in order to produce a psuedorandom sequence of numbers.

PAL   Programmable Array of Logic. A very simple programmable logic device.

PLL   Phase Locked Loop. A device that will output a clock with approximately the same input frequency and phase as an input clock. The PLL will create a regular clock even when the input clock misses cycles. Major changes in the input clock's frequency or phase will be smoothed into gradual changes in the output clock by the PLL.

RTL   Register Transfer Level. The level of coding necessary for synthesis to hardware. This level describes hardware in terms of combinatorial logic and sequential logic dependent on clock edges.

Standard Cell   A specific type of ASIC architecture in which all low-level functions used in the design are placed and routed on the chip.

Synthesis   The process of taking an HDL description of hardware and creating a netlist that can be directly used for manufacturing the hardware.

Index

About the Author

 **Bob Zeidman** is the president of Zeidman Consulting, a hardware and software contract development firm in Silicon Valley, California. Since 1983, he has designed ASICs, FPGAs, and PC boards for RISC-based parallel processor systems, laser printers, communication switches, and other real-time systems. His clients have included Apple Computer, Cisco Systems, Intel, and Texas Instruments. He has written technical papers on hardware and software design methods, and has taught courses on Verilog, ASIC, and FPGA design at engineering conferences throughout the country. Bob is also an instructor and advisor for Semizone.com, an e-learning company for engineers.

Bob's HDL experience began in 1985 at American Supercomputers, where he helped develop a model of a vector supercomputer using the C programming language. Bob was also one of the lead designers of two generations of hardware accelerators for VHDL and Verilog at IKOS Systems, and is often called in by Quickturn Design Systems to develop circuits for use with their hardware emulation systems. He has been involved in a number of projects as a consultant to Cisco Systems including the design of an Ethernet switch and an ATM router, both using large FPGAs designed with Verilog. Also at Cisco, Bob developed the Verilog simulation environment for a large 10-baseT Ethernet switch based on very high density ASICs.

Among Bob's honors are a National Merit Scholarship and a Stanford Graduate Engineering Fellowship for studies in electrical engineering. He also received accolades from the Association for Educational Data Systems, and was the winner of the 1996 American by Design award from Wyle Laboratories and EE Times magazine. He is a Senior Member of the IEEE and his biography is listed in *Who's Who* in America. Bob received his B.A. in Physics and his B.S. in Electrical Engineering from Cornell University, and his M.S. in Electrical Engineering from Stanford University.

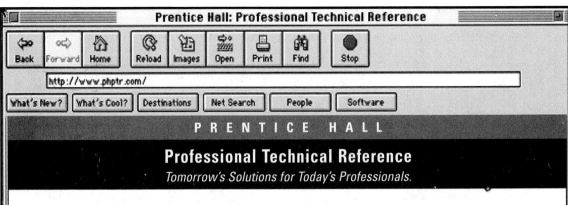

Keep Up-to-Date with
PH PTR Online!

We strive to stay on the cutting-edge of what's happening in professional computer science and engineering. Here's a bit of what you'll find when you stop by **www.phptr.com**:

Special interest areas offering our latest books, book series, software, features of the month, related links and other useful information to help you get the job done.

Deals, deals, deals! Come to our promotions section for the latest bargains offered to you exclusively from our retailers.

Need to find a bookstore? Chances are, there's a bookseller near you that carries a broad selection of PTR titles. Locate a Magnet bookstore near you at www.phptr.com.

What's New at PH PTR? We don't just publish books for the professional community, we're a part of it. Check out our convention schedule, join an author chat, get the latest reviews and press releases on topics of interest to you.

Subscribe Today! **Join PH PTR's monthly email newsletter!**

Want to be kept up-to-date on your area of interest? Choose a targeted category on our website, and we'll keep you informed of the latest PH PTR products, author events, reviews and conferences in your interest area.

Visit our mailroom to subscribe today! **http://www.phptr.com/mail_lists**

LICENSE AGREEMENT AND LIMITED WARRANTY

READ THE FOLLOWING TERMS AND CONDITIONS CAREFULLY BEFORE OPENING THIS CD PACKAGE. THIS LEGAL DOCUMENT IS AN AGREEMENT BETWEEN YOU AND PRENTICE-HALL, INC. (THE "COMPANY"). BY OPENING THIS SEALED CD PACKAGE, YOU ARE AGREEING TO BE BOUND BY THESE TERMS AND CONDITIONS. IF YOU DO NOT AGREE WITH THESE TERMS AND CONDITIONS, DO NOT OPEN THE CD PACKAGE. PROMPTLY RETURN THE UNOPENED CD PACKAGE AND ALL ACCOMPANYING ITEMS TO THE PLACE YOU OBTAINED THEM FOR A FULL REFUND OF ANY SUMS YOU HAVE PAID.

1. **GRANT OF LICENSE:** In consideration of your purchase of this book, and your agreement to abide by the terms and conditions of this Agreement, the Company grants to you a nonexclusive right to use and display the copy of the enclosed software program (hereinafter the "SOFTWARE") on a single computer (i.e., with a single CPU) at a single location so long as you comply with the terms of this Agreement. The Company reserves all rights not expressly granted to you under this Agreement.

2. **OWNERSHIP OF SOFTWARE:** You own only the magnetic or physical media (the enclosed CD) on which the SOFTWARE is recorded or fixed, but the Company and the software developers retain all the rights, title, and ownership to the SOFTWARE recorded on the original CD copy(ies) and all subsequent copies of the SOFTWARE, regardless of the form or media on which the original or other copies may exist. This license is not a sale of the original SOFTWARE or any copy to you.

3. **COPY RESTRICTIONS:** This SOFTWARE and the accompanying printed materials and user manual (the "Documentation") are the subject of copyright. The individual programs on the CD are copyrighted by the authors of each program. Some of the programs on the CD include separate licensing agreements. If you intend to use one of these programs, you must read and follow its accompanying license agreement. You may not copy the Documentation or the SOFTWARE, except that you may make a single copy of the SOFTWARE for backup or archival purposes only. You may be held legally responsible for any copying or copyright infringement which is caused or encouraged by your failure to abide by the terms of this restriction.

4. **USE RESTRICTIONS:** You may not network the SOFTWARE or otherwise use it on more than one computer or computer terminal at the same time. You may physically transfer the SOFTWARE from one computer to another provided that the SOFTWARE is used on only one computer at a time. You may not distribute copies of the SOFTWARE or Documentation to others. You may not reverse engineer, disassemble, decompile, modify, adapt, translate, or create derivative works based on the SOFTWARE or the Documentation without the prior written consent of the Company.

5. **TRANSFER RESTRICTIONS:** The enclosed SOFTWARE is licensed only to you and may not be transferred to any one else without the prior written consent of the Company. Any unauthorized transfer of the SOFTWARE shall result in the immediate termination of this Agreement.

6. **TERMINATION:** This license is effective until terminated. This license will terminate automatically without notice from the Company and become null and void if you fail to comply with any provisions or limitations of this license. Upon termination, you shall destroy the Documentation and all copies of the SOFTWARE. All provisions of this Agreement as to warranties, limitation of liability, remedies or damages, and our ownership rights shall survive termination.

7. **MISCELLANEOUS:** This Agreement shall be construed in accordance with the laws of the United States of America and the State of New York and shall benefit the Company, its affiliates, and assignees.

8. **LIMITED WARRANTY AND DISCLAIMER OF WARRANTY:** The Company warrants that the SOFTWARE, when properly used in accordance with the Documentation, will operate in substantial conformity with the description of the SOFTWARE set forth in the Documentation. The Company does not warrant that the SOFTWARE will meet your requirements or that the operation

of the SOFTWARE will be uninterrupted or error-free. The Company warrants that the media on which the SOFTWARE is delivered shall be free from defects in materials and workmanship under normal use for a period of thirty (30) days from the date of your purchase. Your only remedy and the Company's only obligation under these limited warranties is, at the Company's option, return of the warranted item for a refund of any amounts paid by you or replacement of the item. Any replacement of SOFTWARE or media under the warranties shall not extend the original warranty period. The limited warranty set forth above shall not apply to any SOFTWARE which the Company determines in good faith has been subject to misuse, neglect, improper installation, repair, alteration, or damage by you. EXCEPT FOR THE EXPRESSED WARRANTIES SET FORTH ABOVE, THE COMPANY DISCLAIMS ALL WARRANTIES, EXPRESS OR IMPLIED, INCLUDING WITHOUT LIMITATION, THE IMPLIED WARRANTIES OF MERCHANTABILITY AND FITNESS FOR A PARTICULAR PURPOSE. EXCEPT FOR THE EXPRESS WARRANTY SET FORTH ABOVE, THE COMPANY DOES NOT WARRANT, GUARANTEE, OR MAKE ANY REPRESENTATION REGARDING THE USE OR THE RESULTS OF THE USE OF THE SOFTWARE IN TERMS OF ITS CORRECTNESS, ACCURACY, RELIABILITY, CURRENTNESS, OR OTHERWISE.

IN NO EVENT, SHALL THE COMPANY OR ITS EMPLOYEES, AGENTS, SUPPLIERS, OR CONTRACTORS BE LIABLE FOR ANY INCIDENTAL, INDIRECT, SPECIAL, OR CONSEQUENTIAL DAMAGES ARISING OUT OF OR IN CONNECTION WITH THE LICENSE GRANTED UNDER THIS AGREEMENT, OR FOR LOSS OF USE, LOSS OF DATA, LOSS OF INCOME OR PROFIT, OR OTHER LOSSES, SUSTAINED AS A RESULT OF INJURY TO ANY PERSON, OR LOSS OF OR DAMAGE TO PROPERTY, OR CLAIMS OF THIRD PARTIES, EVEN IF THE COMPANY OR AN AUTHORIZED REPRESENTATIVE OF THE COMPANY HAS BEEN ADVISED OF THE POSSIBILITY OF SUCH DAMAGES. IN NO EVENT SHALL LIABILITY OF THE COMPANY FOR DAMAGES WITH RESPECT TO THE SOFTWARE EXCEED THE AMOUNTS ACTUALLY PAID BY YOU, IF ANY, FOR THE SOFTWARE.

SOME JURISDICTIONS DO NOT ALLOW THE LIMITATION OF IMPLIED WARRANTIES OR LIABILITY FOR INCIDENTAL, INDIRECT, SPECIAL, OR CONSEQUENTIAL DAMAGES, SO THE ABOVE LIMITATIONS MAY NOT ALWAYS APPLY. THE WARRANTIES IN THIS AGREEMENT GIVE YOU SPECIFIC LEGAL RIGHTS AND YOU MAY ALSO HAVE OTHER RIGHTS WHICH VARY IN ACCORDANCE WITH LOCAL LAW.

ACKNOWLEDGMENT

YOU ACKNOWLEDGE THAT YOU HAVE READ THIS AGREEMENT, UNDERSTAND IT, AND AGREE TO BE BOUND BY ITS TERMS AND CONDITIONS. YOU ALSO AGREE THAT THIS AGREEMENT IS THE COMPLETE AND EXCLUSIVE STATEMENT OF THE AGREEMENT BETWEEN YOU AND THE COMPANY AND SUPERSEDES ALL PROPOSALS OR PRIOR AGREEMENTS, ORAL, OR WRITTEN, AND ANY OTHER COMMUNICATIONS BETWEEN YOU AND THE COMPANY OR ANY REPRESENTATIVE OF THE COMPANY RELATING TO THE SUBJECT MATTER OF THIS AGREEMENT.

Should you have any questions concerning this Agreement or if you wish to contact the Company for any reason, please contact in writing at the address below.

Robin Short

Prentice Hall PTR

One Lake Street

Upper Saddle River, New Jersey 07458

About the CD

All code has been simulated using SILOS III version 99.115 from Simucad Inc. All of the RTL code has been synthesized using FPGA Express version 3.4 from Synopsys, Inc. A trial version of the SILOS III simulator is available from Simucad at their website www.simucad.com. A trial version of the FPGA Compiler II synthesis tool is available from Synopsys at their website www.synopsys.com.

Technical Support

Prentice-Hall does not offer technical support for this software. If there is a problem with the media, however, you may obtain a replacement CD by emailing a description of the problem. Send your email to: disc_exchange@prenhall.com.